AF576025

Nanomaterials and Nanofabrication for Electrochemical Energy Storage

Nanomaterials and Nanofabrication for Electrochemical Energy Storage

Special Issue Editors

Jian Liu
Dongping Lu
Xiaolei Wang

MDPI • Basel • Beijing • Wuhan • Barcelona • Belgrade • Manchester • Tokyo • Cluj • Tianjin

Special Issue Editors
Jian Liu
University of British Columbia
Canada

Dongping Lu
Pacific Northwest National Laboratory
USA

Xiaolei Wang
University of Alberta
Canada

Editorial Office
MDPI
St. Alban-Anlage 66
4052 Basel, Switzerland

This is a reprint of articles from the Special Issue published online in the open access journal *Nanomaterials* (ISSN 2079-4991) (available at: https://www.mdpi.com/journal/nanomaterials/special_issues/electroces_nano).

For citation purposes, cite each article independently as indicated on the article page online and as indicated below:

LastName, A.A.; LastName, B.B.; LastName, C.C. Article Title. *Journal Name* **Year**, *Article Number*, Page Range.

ISBN 978-3-03936-278-3 (Hbk)
ISBN 978-3-03936-279-0 (PDF)

Contents

About the Special Issue Editors

Jian Liu is an Assistant Professor at the University of British Columbia (UBC) Okanagan campus, Canada. Dr. Liu received his Ph.D. in materials science in 2013 from the University of Western Ontario (Canada), and worked as a NSERC Postdoctoral Fellow at Lawrence Berkeley National Laboratory and Pacific Northwest National Laboratory (USA) prior to joining UBC in January 2017. Dr. Liu is leading the Advanced Battery Facility at UBC (http://nesc.ok.ubc.ca/), and his current research interests focus on advanced nanofabrication techniques, materials design for Li-ion batteries and beyond, and interfacial control and understanding in energy storage systems. Dr. Liu has published over 70 research papers in high-quality peer-reviewed journals, such as *Nature Communication, Nature Energy, Advanced Materials, Nano Letters*, etc. Dr. Liu is the recipient of many prestigious awards, including the NSERC PDF Award, PNNL Alternated-Sponsored PDF, Mitacs Elevate PDF, and UBC Emerging Professor Award.

Dongping Lu is a materials scientist and the team leader of the Battery Materials Research team in the Battery Materials and Systems group at Pacific Northwest National Laboratory (PNNL). Dr. Lu's research interest is material and electrolyte development for energy storage as well as in-depth mechanism analysis through advanced in-situ/ex-situ techniques. Currently, he is focusing on the development of high energy lithium-sulfur and solid batteries including new materials, fundamental understanding, and pouch cell design and demonstration. He received his Ph.D. in Physical Chemistry (Electrochemistry) at Xiamen University, China in 2011. Dr. Lu has more than 40 papers published in peer-reviewed professional journals and holds 6 granted and pending U.S. Patents.

Xiaolei Wang is a tenure-track assistant professor of chemical engineering at the University of Alberta (UofA) with expertise in advanced materials, nanotechnology, and clean energy technologies. His research themes centre upon the design, development, and application of novel nanostructured materials for energy-related technologies including lithium-ion batteries, sodium (and other alkaline)-ion batteries, lithium-sulfur (Li-S) batteries, aqueous batteries, and electrocatalytic systems such as metal-air batteries, water electrolyzers, and systems for electrochemical CO_2 reduction. Before joining UofA, he worked as a tenure-track assistant professor at Concordia University from 2017 to 2019, and postdoctoral fellow researcher at the University of Waterloo from 2014 to 2017. Dr. Wang received his Ph.D. in Chemical and Biomolecular Engineering at the University of California, Los Angeles (UCLA, 2013), M.Sc. in Chemical Engineering at Tianjin University (TJU, 2007), and B.Sc. in Polymer Chemical Engineering at Dalian University of Technology (DUT, 2004). Dr. Wang builds upon the knowledge from the fundamental studies and understanding of mechanisms by correlating the electrochemical performance of clean energy technologies with materials' morphologies and microstructures, aiming to develop next-generation high-performance clean energy technologies for practical applications. He has published more than 50 articles in peer-reviewed journals and received more than 3800 citations with an h-index of 31 (May 2020). His work is recognized by numerous awards, including NSERC Discovery Accelerator (2018), Petro-Canada Young Investigator Award (2018) and Concordia University Research Chair-Young Scholar (2019).

Preface to "Nanomaterials and Nanofabrication for Electrochemical Energy Storage"

Electrochemical energy storage technologies play key roles for storing electricity harvested from renewable energy resources of an intermittent nature, such as solar and wind, and for utilizing electricity for a range of applications, such as electric vehicles and flights, wearable electronics, and medical implants. Several electrochemical systems, such as rechargeable batteries and supercapacitors, have shown great potential for these emerging applications. In these systems, nanostructured materials have been widely used for improving electrochemical performance, and for studying electrochemical reaction mechanisms due to their unique chemical and physical properties. For these applications, it is essential to design and synthesize novel multiscale nanomaterials with optimized structure and properties by using nanofabrication techniques.

The papers collected in this Special Issue include original research from controlled synthesis of various nanomaterials (porous carbon, graphene, solid electrolyte, thin film, one-dimensional nanowires and nanotubes); advanced characterizations; and applications in Li-ion batteries, supercapacitors, and Zn-ion batteries. The Guest Editors are thankful to all the authors who contributed their excellent works to this Special Issue. We hope researchers in the field will benefit from the results published herein for their future work.

Jian Liu, Dongping Lu, Xiaolei Wang
Special Issue Editors

Article

Spark Plasma Sintering of Lithium Aluminum Germanium Phosphate Solid Electrolyte and Its Electrochemical Properties

Hongzheng Zhu, Anil Prasad, Somi Doja, Lukas Bichler and Jian Liu *

School of Engineering, Faculty of Applied Science, The University of British Columbia, Kelowna, BC V1V 1V7, Canada
* Correspondence: Jian.liu@ubc.ca

Received: 14 July 2019; Accepted: 26 July 2019; Published: 29 July 2019

Abstract: Sodium superionic conductor (NASICON)-type lithium aluminum germanium phosphate (LAGP) has attracted increasing attention as a solid electrolyte for all-solid-state lithium-ion batteries (ASSLIBs), due to the good ionic conductivity and highly stable interface with Li metal. However, it still remains challenging to achieve high density and good ionic conductivity in LAGP pellets by using conventional sintering methods, because they required high temperatures (>800 °C) and long sintering time (>6 h), which could cause the loss of lithium, the formation of impurity phases, and thus the reduction of ionic conductivity. Herein, we report the utilization of a spark plasma sintering (SPS) method to synthesize LAGP pellets with a density of 3.477 g cm^{-3}, a relative high density up to 97.6%, and a good ionic conductivity of 3.29×10^{-4} S cm^{-1}. In contrast to the dry-pressing process followed with high-temperature annealing, the optimized SPS process only required a low operating temperature of 650 °C and short sintering time of 10 min. Despite the least energy and short time consumption, the SPS approach could still achieve LAGP pellets with high density, little voids and cracks, intimate grain–grain boundary, and high ionic conductivity. These advantages suggest the great potential of SPS as a fabrication technique for preparing solid electrolytes and composite electrodes used in ASSLIBs.

Keywords: spark plasma sintering; NASICON-type; ionic conductivity; solid electrolyte; solid–solid interface; grain-boundary resistance

1. Introduction

The utilization of rechargeable lithium-ion batteries (LIBs) has been increasingly expanded from consumer devices to plug-in or hybrid electric vehicles and smart grid energy storage systems, in order to alleviate the dependence on fossil fuels, decrease greenhouse gas emissions, and realize clean transportation [1,2]. However, traditional LIBs utilize liquid organic electrolytes that could cause catastrophic disasters (such as fire and explosion) upon exposure to the air due to their flammable and volatile nature [3–5]. Moreover, the energy densities of current LIBs are gradually approaching their theoretical limits and cannot meet the increasing demands from end users for higher-performance LIBs. From this aspect, all-solid-state lithium-ion batteries (ASSLIBs) have recently been revisited as promising next-generation battery systems because of their significantly improved energy density and safety [3,6–8]. The major difference between ASSLIBs and conventional LIBs is the use of solid-state electrolytes (SSEs), rather than liquid organic electrolytes, in the design and fabrication of battery systems. The adoption of SSEs not only can significantly reduce the safety risks associated with the flammable, volatile, and toxic liquid organic electrolytes, but also could potentially address the Li dendrite growth problem and make it possible for the utilization of Li metal with a high theoretical capacity of 3860 mA h g^{-1} as the anode, therefore substantially elevating energy densities and the safety of ASSLIBs.

Among different kinds of SSEs, including garnet [9,10], perovskite [11,12], sulfides [13], hydrides [14], borate or phosphate [15–17], halides [18], lithium phosphorousoxynitride (LiPON) [19], lithium superionic conductor (LISICON) [20–23], and other Li-based ceramic [24], sodium superionic conductor (NASICON)-type is one of the most popular solid electrolytes, due to its high ionic conductivity and good chemical and thermal stability with lithium anode [25–28]. In particular, lithium aluminum germanium phosphate (LAGP) has a NASICON-type structure and possesses several advantages for material preparation and practical applications in ASSLIBs. First, the high stability of LAGP SSEs against O_2 and H_2O means that the synthesis of materials and the assembly of batteries could be performed in an ambient atmosphere, therefore simplifying the manufacturing processes and requirements [3–5]. Secondly, the predominant ionic conductivities of LAGP SSEs are in the order of 10^{-3}–10^{-5} S cm^{-1} at room temperature (RT), which are relatively high compared with other ceramic electrolytes [29,30]. Thirdly, LAGP SSEs exhibit a large electrochemical stability window (1.8–7 V vs. Li^+/Li), good chemical compatibility with cathode materials at different charge states, and excellent interfacial stability towards Li metal anode [6,31,32]. Nevertheless, it was also reported that short-circuit occurred in SSEs because the deposited lithium grew along voids among grains, and an effective approach to suppress lithium dendrites was to enhance the density and the mechanical strength of SSEs and create stable grain–grain interfaces in the solid electrolytes [33]. Therefore, it is imperative to develop methods to fabricate SSE pellets with high density and free of voids and through-holes. However, conventional sintering methods usually require high temperatures (>800 °C) and long sintering time (>6 h), which could easily result in grain coarsening, formation of impurity phases, and thus reduced ionic conductivity [34].

In recent years, spark plasma sintering (SPS) is attracting increasing attention from researchers to fabricate solid electrolytes and electrodes with the densities close to their theoretical values in a short sintering time [35,36]. During SPS processes, uniaxial force and pulsed direct electrical current are simultaneously applied to the powders and thus can rapidly consolidate the powders into dense pellets. The use of microscopic electrical current allows the sintering to complete in a few minutes, because of the very high heating rates and the localized high temperatures between particles. In the meantime, the water-cooling system in SPS allows a very high cooling rate. Compared to the conventional heat treatment, the effective heating and cooling systems in SPS enhanced the densification of SSE powders through grain diffusion mechanisms and avoided grain coarsening to maintain the intrinsic merits of nano-powders [37,38]. Indeed, SPS technique, with the advantages of a flash and short processing time, improves the sintering ability of various powder materials and creates intimate solid–solid interfaces in solid electrolytes and electrodes for ASSLIBs [30,39].

In this work, we have successfully produced high-densified LAGP pellets, with a high ionic conductivity of 3.29 × 10^{-4} S cm^{-1} by using SPS. Owing to the advantage of minimum energy and time consumptions of SPS, the highly dense LAGP without microcracking resulted in the reduced resistance at grain boundaries, due to the removal of the pores/voids/cracks at a proper sintering temperature, thus improving the overall ionic conductivity of LAGP. At the same time, the low operating SPS temperature avoided the formation of ionic nonconductive impurities, which usually appeared in the grain boundaries for traditionally high-temperature sintered samples and resulted in blocking of Li-ion transport pathways.

2. Materials and Methods

LAGP powders with a stoichiometric formula of $Li_{1.5}Al_{0.5}Ge_{1.5}(PO_4)_3$ were placed in a graphite tooling (1 mm die) of the SPS chamber (Thermal Technologies 10-3 SPS system) and heated up to the target sintering temperature in argon atmosphere. The SPS sintering pressure was 60 MPa, the sintering temperatures were changed from 600 °C to 750 °C, and the sintering time was varied between 1 min and 10 min for all the experiments. The SPS heating rate was set to 100 °C min^{-1}. Upon the completion of SPS sintering, the samples were cooled down to room temperature (RT) naturally. In addition to the SPS, LAGP pellets were also made by a conventional dry-pressing method, where 0.5 g of LAGP powder was placed into a stainless steel die with 10 mm diameter and pressed at 200 MPa for 2 min. The obtained pellets were then placed into a tube furnace and annealed at 800 °C for 6 h under nitrogen atmosphere.

The phase structure of the starting LAGP powders and the LAGP pellets were characterized by X-ray diffraction (XRD, D8-Advance X-ray diffractometer, Bruker, Karlsruhe, Germany) with Cu $K_{\alpha 1}$ and $K_{\alpha 2}$ radiation source. The LAGP microstructure was observed by field-emission scanning electron microscopy (MIRA3 FEG-SEM, Tescan, Brno, Czech Republic). The density of the LAGP pellets was measured by the Archimedes' method. For ionic conductivity measurement, the surfaces of the LAGP pellets were first polished with sandpapers to remove carbon residual from the graphite die during the SPS process, and then coated with 100 nm Au layers on both sides as blocking electrodes. An electrochemical impedance spectroscopy (EIS) test of the Au-coated LAGP pellets was conducted on a Potentiostat/Galvanostat Station (Biologic VSP) in a frequency range of 0.1 Hz–1 MHz at different temperatures of −10 °C to 80 °C.

3. Results and Discussion

As the starting material, LAGP powders were characterized by SEM for the morphology and by XRD for the phase structure, and the results are presented in Figure 1. As shown in Figure 1a, the diameter of LAGP particles was about 400–800 nm. The XRD pattern of LAGP powders (Figure 1b) shows strong diffraction peaks, which could be well indexed to $LiGe_2(PO_4)_3$ with a NASICON-type structure (JCPDS PDF No. 80-1922). No other peaks of impurities are observed in the XRD pattern, indicating the high purity of the starting LAGP powders.

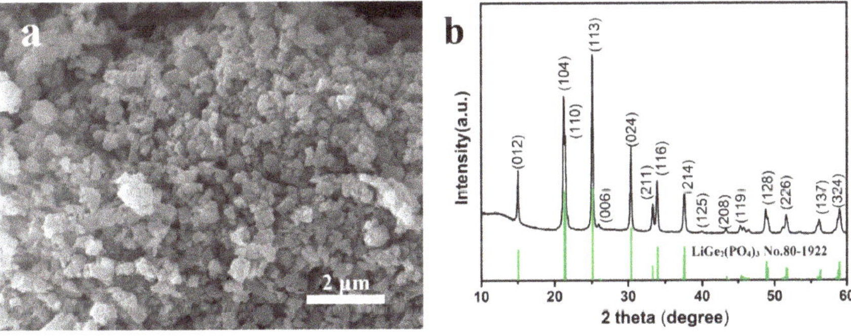

Figure 1. (**a**) SEM image and (**b**) XRD pattern of lithium aluminum germanium phosphate (LAGP) powders.

The effects of SPS sintering temperatures (600–750 °C) on the morphology, ionic conductivity, and phase structure of LAGP pellets was studied first. Figure 2a–d presents the SEM images of the cross-sectional view of LAGP pellets sintered by SPS at different temperatures of 600 °C, 650 °C, 700 °C, and 750 °C for 2 min. It can be observed that the grain size in all the LAGP pellets was below 800 nm, similar to that of the starting LAGP powders (Figure 1a). The LAGP grain size remained almost unchanged at sintering temperatures of 600–750 °C, due to short SPS sintering time (2 min). As seen in Figure 2a,b, LAGP pellets fabricated at 600 °C and 650 °C were dense with minimal visible porosity. In contrast, the sample sintered at 700 °C in Figure 2c exhibited enlarging pores at grain boundaries, and the one sintered at 750 °C in Figure 2d possessed large cracks in the pellets' cross section, as seen in Figure S1. The high temperature at 750 °C possibly resulted in rapid grain growth, leading to the relatively high thermal expansion anisotropy of LAGP. Thus, the growing pores and voids likely coalesced and gradually emerged on the sample's surfaces.

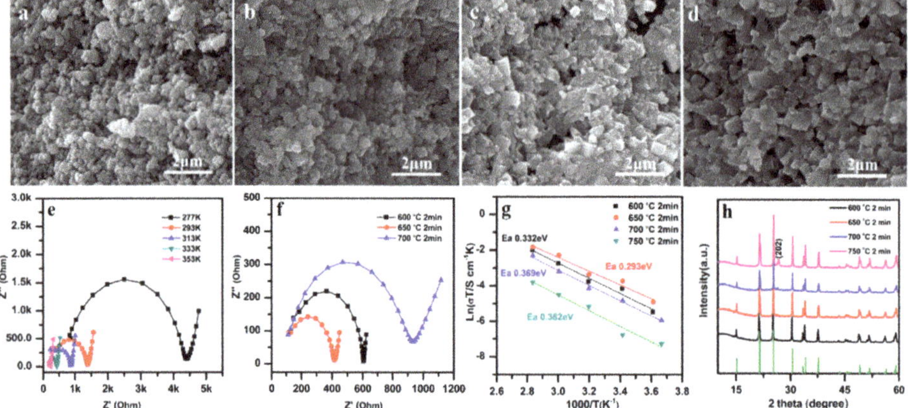

Figure 2. SEM images at cross sections of the LAGP pellets sintered by spark plasma sintering (SPS) for 2 min at (**a**) 600 °C, (**b**) 650 °C, (**c**) 700 °C, and (**d**) 750 °C; (**e**) Nyquist plots of the LAGP pellet, sintered at 650 °C for 2 min; (**f**) Nyquist plots at room temperature (RT), (**g**) Arrhenius curves, and (**h**) XRD patterns of LAGP pellets, sintered by SPS for 2 min at 600 °C, 650 °C, 700 °C, and 750 °C.

The Nyquist plots of the LAGP pellets were measured by EIS and fitted by using an equivalent circuit of $R_1(R_2||\text{CPE})W$ to obtain their ionic conductivities. As illustrated in Figure 2e,f, a typical Nyquist plot consists of one semicircle in the high-frequency region and a large spike in the low-frequency region. In general, the intercept of the high-frequency semicircle with the real axis stands for the total resistance of $R_2 = (R_{\text{bulk}} + R_{\text{interface}})$, where R_{bulk} and $R_{\text{interface}}$ represent the bulk resistance and grain-interface resistance, respectively. The spike in the low-frequency domain is known as Warburg diffusion impedance, which is due to the polarization resulting from Li-ion conduction at the electrolyte/Au electrode interface [40]. Figure 2e shows one EIS example of LAGP sintered at 650 °C, measured at different temperatures between −10 °C and 80 °C. The complex Nyquist plots of the LAGP pellets prepared between 600 °C and 750 °C are shown in Figure 2f. As supported by the SEM images (Figure 2c,d, and Figure S1), the appearance of voids and cracks in the LAGP pellets sintered at high temperatures (700 °C and 750 °C) likely caused the increase in the resistance and the reduction in ionic conductivity.

The values of ionic conductivities can be determined from $\sigma = d/AR$, in which d and A stand for the thickness and the area of the pellet, respectively, and R is the total resistance (R_2), obtained above, and the results are given in Figure 2g. Among all the samples, the LAGP pellets sintered at 650 °C for 2 min exhibited the highest conductivity of 8.09×10^{-5} S cm^{-1} at RT. The plots of $log(\sigma T)$ vs. $1000/T$ show a linear relation and fit well with the Arrhenius equation, $\sigma = A\ exp\ (E_a/kT)$, in which A, E_a, and k represents the pre-exponential factor, the activation energy for conduction, and the Boltzmann constant, respectively. As seen from the Arrhenius plots (Figure 2g), the slope of the 650 °C sample is the lowest. The activation energies (E_a) are calculated from the rates of slopes, and the 650 °C sample's E_a is 0.293 eV. In Figure 2h, the XRD patterns of SPS pellets sintered at different temperatures are given. All of the LAGP pellets sintered at 600 °C to 700 °C maintained pure NASICON-type phase (JCPDS PDF No. 80-1922). When the temperature increased to 750 °C, the peak intensity increased at 26.4°, which corresponds to the (202) crystal plane of NASICON-type phase. This might be due to the reorientation of LAGP grains caused by the high pressure during the SPS process. For the whole heat treatment, the width of the diffraction peaks did not change, implying that the average crystallite size of LAGP remained unchanged.

LAGP SSEs sintered by SPS method could usually reach a very high density in a relatively low temperature (600–750 °C) and short time (2 min). The density of LAGP SSEs sintered at 600 °C was 3.219 g cm^{-3}, as seen in Table 1. When the sintering temperature increased to 650 °C, the density increased up to 3.477 g cm^{-3}. However, when the temperature further elevated to 700 °C and 750 °C, the density decreased to 2.495 and 2.430 g cm^{-3}, respectively, most likely due to the formation of voids and microcracks in the pellets. The relatively density of LAGP SSEs was determined to be 90.4%, 97.6%, 70.1%, and 68.2%, for the LAGP pellets sintered at 600 °C, 650 °C, 700 °C, and 750 °C, respectively. Therefore, the SPS sintering temperature is a key parameter to determine the density and ionic conductivity of LAGP pellets.

Table 1. Comparison of the measured densities and the relative densities of LAGP sintered by SPS method at different temperatures.

SPS Sintering Temperature (°C)	**600**	**650**	**700**	**650**
Measured density (g cm^{-3})	3.219	3.477	2.495	2.430
Relative density (%) *	90.4	97.6	70.1	68.2

* Note: Relative density = measured density/theoretical density × 100. LAGP has a theoretical density of 3.5615 g cm^{-3}.

The SEM images of LAGP pellets subjected to 650 °C SPS sintering temperature for different times from 1 min, 2 min, 5 min, to 10 min are shown in Figure 3a–d, respectively. The microstructure of LAGP started to reorganize after 5 min of sintering, due to the melting that occurred at grain boundaries (Figure 3c). Most of the LAGP particles were fully joined after 10 min of sintering, and a highly densified structure was finally obtained (Figure 3d). SPS has a totally different densification mechanism compared with that of a conventional heating treatment. During the SPS process, Joule heating was introduced at the physical contact points of different particles, causing localized heating and possibly melting to facilitate the densification of the LAGP powders [41].

Figure 3. SEM images at cross sections of the LAGP pellets sintered at 650 °C for different sintering times: (**a**) 1 min, (**b**) 2 min, (**c**) 5 min, and (**d**) 10 min; (**e**) Nyquist plots of LAGP pellets (650 °C, 10 min) measured at different temperatures; (**f**) Nyquist plots, (**g**) Arrhenius plots, and (**h**) XRD patterns of the LAGP fabricated at 650 °C for 2 min and 10 min.

The EIS results for the samples obtained at 650 °C for 10 min are presented in Figure 3e, and the Nyquist plots of the samples sintered at 650 °C with different SPS times are shown in Figure 3f. It is obvious that the resistance of LAGP pellets decreased as the sintering time increased, as reflected by the reduction of the high-frequency semicircles in Figure 3f. When the sintering time reached 10 min, the LAGP pellets exhibited the highest ionic conductivity of 3.29×10^{-4} S cm^{-1} at RT. This result is consistent with the SEM microstructure observation (Figure 3d), which discloses the melting of grain boundaries. The high densification and well-jointed grain boundaries reduced the total resistance of the LAGP pellets. The E_a calculation of the SPS samples processed at different times reveals that the 650 °C 10 min sample had a minimum E_a of 0.239 eV, suggesting the lowest diffusion barrier for Li ions in SSEs. No impurity phases were detected in the XRD pattern for LAGP pellets sintered at 650 °C for 10 min.

In order to validate our sintering approach, it is necessary to investigate the impacts of the SPS compared with other conventional sintering methods. Therefore, LAGP pellets were also synthesized by a dry-pressing method, followed by heat treatment at 800 °C for 6 h, which was an optimal condition reported in a previous study [42]. Figure 4a,c show the SEM images of the samples obtained using a conventional sintering approach. It can be seen that the grain size of conventionally sintered LAGP pellet was in a range of 1.0–1.5 μm, which was much larger than that of SPS-sintered LAGP pellets (0.4–0.8 μm), as shown in Figure 4b,d.

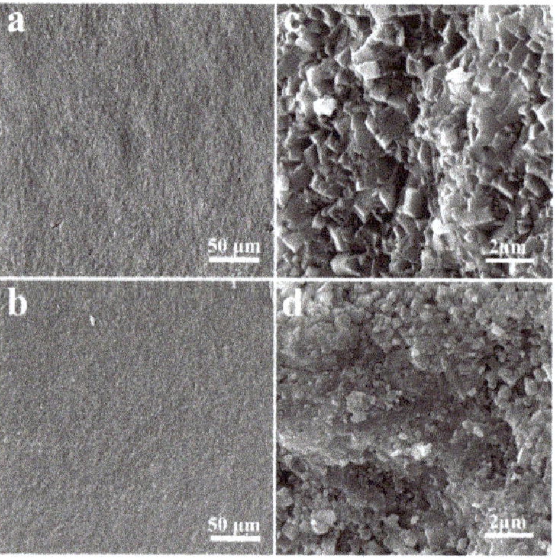

Figure 4. Cross-sectional SEM images of (**a**,**c**) conventionally sintered LAGP pellets and (**b**,**d**) SPS-sintered LAGP pellets.

Table 2 summarizes the correlations of preparation methods and conditions with the final density, relative density, ionic conductivity, and activation energy for LAGP SSEs. In conventional sintering approaches, high-temperature heat treatment is usually required to improve grain–grain interfaces and minimize grain-boundary resistance [31,42–45]. While low-temperature annealing leads to a low ionic conductivity, such as the glassy LAGP sample annealed at 500 °C, which is lower than its crystallization temperature [38]. This proved that the glassy phase of LAGP is not as good a Li-ion conductor as the crystallized one. However, two detrimental effects, i.e., the loss of lithium and formation of second phases, are usually concurrent for high-temperature heat treatment, and could cause the reduction of ionic conductivity in LAGP SSEs [31,42–45]. In our optimal SPS process, the sintering temperature was only 650 °C, which is much lower than other reported sintering methods, and yet the ionic conductivity remained high. At last but not least, LAGP pellets fabricated by SPS possess the highest relative density of 97.6% (Table 2).

Table 2. Summary of the density, relatively density, ionic conductivity, and activation energy for LAGP pellets fabricated using different methods.

SSEs	Preparation Method	Preparation Condition	Density (g cm^{-3})	Relative Density (%)	Ionic Conductivity (S cm^{-1})	Activation Energy (eV)	Ref.
$Li_{1.5}Al_{0.5}Ge_{1.5}(PO_4)_3$	hot press + post-calcination	600 °C, 1 h, 200 MPa + 800 °C, 5 h	3.33	93.5	1.64×10^{-4}	0.299	[31]
$Li_{1.3}Al_{0.3}Ge_{1.7}(PO_4)_3$	dry-pressing + post-calcination	5 min, 20 MPa + 750 °C, 5 h	3.18	89.3	3.4×10^{-4}	0.330	[43]
$Li_{1.70}Al_{0.61}Ge_{1.35}P_{3.04}O_{12.0}$	cold press + sintering	300 °C, 6 h	3.02	84.8	2.3×10^{-4}	0.320	[42]
$Li_{1.31}Al_{0.42}Ge_{1.52}P_{3.09}O_{12.1}$	quenched glass piece + sintering	500 °C, 1 h	3.08	86.5	Too low to be measured	N/A	[42]
$Li_{1.5}Al_{0.5}Ge_{1.5}(PO_4)_3$	melt-quench + post crystallization	300 °C, 8 h	N/A	N/A	5×10^{-4}	0.280	[44]
$Li_{1.5}Al_{0.5}Ge_{1.5}(PO_4)_3$	cold press + sintering	200 MPa + 850 °C, 5 h	N/A	N/A	2.99×10^{-4}	N/A	[45]
$Li_{1.5}Al_{0.5}Ge_{1.5}(PO_4)_3$	SPS	650 °C 10 min, 100 MPa	3.477	97.6	3.29×10^{-4}	0.239	This work

N/A: not available.

4. Conclusions

In summary, highly densified LAGP pellets were fabricated by using SPS technique and exhibited high ionic conductivity at room temperature. The influence of SPS sintering temperatures (600–750 °C) and times (1–10 min) on the microstructure, density, and ionic conductivity of the LAGP pellets were studied in detail. It can be concluded that the optimal SPS condition for LAGP pellets was 650 °C and 10 min, and the synthesized LAGP pellets were dense with minimal visible porosity. The density of the LAGP pellets was as high as 3.477 g cm^{-3}, a relative density of 97.6% and the ionic conductivity was 3.29×10^{-4} S cm^{-1}, with a activation energy of 0.239 eV at RT. Sintering temperatures higher than 650 °C caused the formation of voids and microcracks in the LAGP pellets, negatively affecting the density and ionic conductivity. These results clearly demonstrate the potential of the SPS approach to synthesize solid electrolytes used in ASSLIBs, owing to the drastically reduced sintering temperature and time compared with that used in conventional sintering processes.

Supplementary Materials: The following are available online at http://www.mdpi.com/2079-4991/9/8/1086/s1, Figure S1: Low-magnification SEM images of SPS pellets sintered for 2 min at different temperatures of (a) 650 °C, (b) 700 °C, and (c) 750 °C.

Author Contributions: Conceptualization, H.Z. and J.L.; Funding acquisition, J.L.; Investigation, H.Z.; Methodology, A.P. and S.D.; Project administration, J.L.; Resources, L.B.; Supervision, J.L.; Writing – original draft, H.Z.; Writing – review & editing, A.P., S.D., L.B. and J.L.

Acknowledgments: This work was supported by the Nature Sciences and Engineering Research Council of Canada (NSERC), Canada Foundation for Innovation (CFI), B.C. Knowledge Development Fund (BCKDF), and the University of British Columbia (UBC). Hongzheng Zhu is grateful for financial support from Chinese Scholarship Council (CSC).

Conflicts of Interest: The authors declare no conflict of interest.

References

1. Guan, C.; Wang, J. Recent development of advanced electrode materials by atomic layer deposition for electrochemical energy storage. *Adv. Sci.* **2016**, *3*, 1500405. [CrossRef]
2. Kim, S.; Cho, W.; Zhang, X.; Oshima, Y.; Choi, J.W. A stable lithium-rich surface structure for lithium-rich layered cathode materials. *Nat. Commun.* **2016**, *7*, 13598. [CrossRef]
3. Manthiram, A. An outlook on lithium ion battery technology. *ACS Cent. Sci.* **2017**, 3, 1063–1069. [CrossRef]
4. Abada, S.; Marlair, G.; Lecocq, A.; Petit, M.; Sauvant-Moynot, V.; Huet, F. Safety focused modeling of lithium-ion batteries: A review. *J. Power Sources* **2016**, *306*, 178–192. [CrossRef]
5. Deng, Z.; Mo, Y.; Ong, S.P. Computational studies of solid-state alkali conduction in rechargeable alkali-ion batteries. *NPG Asia Mater.* **2016**, *8*, e254. [CrossRef]
6. Weiss, M.; Weber, D.A.; Senyshyn, A.; Janek, J.; Zeier, W.G. Correlating transport and structural properties in $Li_{1+x}Al_xGe_{2-x}(PO_4)_{(3)}$ (LAGP) prepared from aqueous solution. *ACS Appl. Mater. Interfaces* **2018**, *10*, 10935–10944. [CrossRef]
7. Sun, C.; Huang, X.; Jin, J.; Lu, Y.; Wang, Q.; Yang, J.; Wen, Z. An ion-conductive $Li_{1.5}Al_{0.5}Ge_{1.5}(PO_4)_{(3)}$-based composite protective layer for lithium metal anode in lithium-sulfur batteries. *J. Power Sources* **2018**, *377*, 36–43. [CrossRef]
8. Jung, Y.-C.; Park, M.-S.; Doh, C.-H.; Kim, D.-W. Organic-inorganic hybrid solid electrolytes for solid-state lithium cells operating at room temperature. *Electrochim. Acta* **2016**, *218*, 271–277. [CrossRef]
9. Kazyak, E.; Chen, K.-H.; Wood, K.N.; Davis, A.L.; Thompson, T.; Bielinski, A.R.; Sanchez, A.J.; Wang, X.; Wang, C.; Sakamoto, J.; et al. Atomic layer deposition of the solid electrolyte garnet $Li_7La_3Zr_2O_{12}$. *Chem. Mater.* **2017**, *29*, 3785–3792. [CrossRef]
10. Dong, Z.; Xu, C.; Wu, Y.; Tang, W.; Song, S.; Yao, J.; Huang, Z.; Wen, Z.; Lu, L.; Hu, N. Dual substitution and spark plasma sintering to improve ionic conductivity of garnet $Li_7La_3Zr_2O_{12}$. *Nanomaterials* **2019**, *9*, 721. [CrossRef]
11. Nguyen, H.; Hy, S.; Wu, E.; Deng, Z.; Samiee, M.; Yersak, T.; Luo, J.; Ong, S.P.; Meng, Y.S. Experimental and computational evaluation of a sodium-rich anti-perovskite for solid state electrolytes. *J. Electrochem. Soc.* **2016**, *163*, A2165–A2171. [CrossRef]

12. Jay, E.E.; Rushton, M.J.D.; Chroneos, A.; Grimes, R.W.; Kilner, J.A. Genetics of superionic conductivity in lithium lanthanum titanates. *Phys. Chem. Chem. Phys.* **2015**, *17*, 178–183. [CrossRef]
13. Kim, K.-B.; Dunlap, N.A.; Han, S.S.; Jeong, J.J.; Kim, S.C.; Oh, K.H.; Lee, S.-H. Nanostructured Si/C fibers as a highly reversible anode material for all-solid-state lithium-ion batteries. *J. Electrochem. Soc.* **2018**, *165*, A1903–A1908. [CrossRef]
14. Unemoto, A.; Yoshida, K.; Ikeshoji, T.; Orimo, S.-I. Bulk-type all-solid-state lithium batteries using complex hydrides containing cluster-anions. *Mater. Trans.* **2016**, *57*, 1639–1644. [CrossRef]
15. Kaneko, F.; Wada, S.; Nakayama, M.; Wakihara, M.; Koki, J.; Kuroki, S. Capacity fading mechanism in all solid-state lithium polymer secondary batteries using PEG-borate/aluminate ester as plasticizer for polymer electrolytes. *Adv. Funct. Mater.* **2009**, *19*, 918–925. [CrossRef]
16. Masuda, Y.; Nakayama, M.; Wakihara, M. Fabrication of all solid-state lithium polymer secondary batteries using PEG-borate/aluminate ester as plasticizer for polymer electrolyte. *Solid State Ionics* **2007**, *178*, 981–986. [CrossRef]
17. Lin, Z.; Liu, Z.; Fu, W.; Dudney, N.J.; Liang, C. Lithium polysulfidophosphates: A family of lithium-conducting sulfur-rich compounds for lithium-sulfur batteries. *Angew. Chem. Int. Ed.* **2013**, *52*, 7460–7463. [CrossRef]
18. Asano, T.; Sakai, A.; Ouchi, S.; Sakaida, M.; Miyazaki, A.; Hasegawa, S. Solid halide electrolytes with high lithium-ion conductivity for application in 4 V class bulk-type all-solid-state batteries. *Adv. Mater.* **2018**, *30*. [CrossRef]
19. Jadhav, H.S.; Kalubarme, R.S.; Jadhav, A.H.; Seo, J.G. Highly stable bilayer of LiPON and B_2O_3 added $Li_{1.5}Al_{0.5}Ge_{1.5}(PO_4)$ solid electrolytes for non-aqueous rechargeable Li-O_2 batteries. *Electrochim. Acta* **2016**, *199*, 126–132. [CrossRef]
20. Ulissi, U.; Agostini, M.; Ito, S.; Aihara, Y.; Hassoun, J. All solid-state battery using layered oxide cathode, lithium-carbon composite anode and thio-LISICON electrolyte. *Solid State Ionics* **2016**, *296*, 13–17. [CrossRef]
21. Nagao, M.; Imade, Y.; Narisawa, H.; Kobayashi, T.; Watanabe, R.; Yokoi, T.; Tatsumi, T.; Kanno, R. All-solid-state Li-sulfur batteries with mesoporous electrode and thio-LISICON solid electrolyte. *J. Power Sources* **2013**, *222*, 237–242. [CrossRef]
22. Kobayashi, T.; Yamada, A.; Kanno, R. Interfacial reactions at electrode/electrolyte boundary in all solid-state lithium battery using inorganic solid electrolyte, thio-LISICON. *Electrochim. Acta* **2008**, *53*, 5045–5050. [CrossRef]
23. Kobayashi, T.; Imade, Y.; Shishihara, D.; Homma, K.; Nagao, M.; Watanabe, R.; Yokoi, T.; Yamada, A.; Kanno, R.; Tatsumi, T. All solid-state battery with sulfur electrode and thio-LISICON electrolyte. *J. Power Sources* **2008**, *182*, 621–625. [CrossRef]
24. Kuganathan, N.; Tsoukalas, L.H.; Chroneos, A. Defects, dopants and Li-ion diffusion in Li_2SiO_3. *Solid State Ionics* **2019**, *335*, 61–66. [CrossRef]
25. Yu, R.; Du, Q.-X.; Zou, B.-K.; Wen, Z.-Y.; Chen, C.-H. Synthesis and characterization of perovskite-type (Li,Sr)(Zr,Nb)O-3 quaternary solid electrolyte for all-solid-state batteries. *J. Power Sources* **2016**, *306*, 623–629. [CrossRef]
26. Lalere, F.; Leriche, J.B.; Courty, M.; Boulineau, S.; Viallet, V.; Masquelier, C.; Seznec, V. An all-solid state NASICON sodium battery operating at 200 degrees C. *J. Power Sources* **2014**, *247*, 975–980. [CrossRef]
27. Noguchi, Y.; Kobayashi, E.; Plashnitsa, L.S.; Okada, S.; Yamaki, J.-i. Fabrication and performances of all solid-state symmetric sodium battery based on NASICON-related compounds. *Electrochim. Acta* **2013**, *101*, 59–65. [CrossRef]
28. Mahmoud, M.M.; Cui, Y.; Rohde, M.; Ziebert, C.; Link, G.; Seifert, H.J. Microwave crystallization of lithium aluminum germanium phosphate solid-state electrolyte. *Mater.* **2016**, *9*, 506. [CrossRef]
29. Liu, Y.; Li, C.; Li, B.; Song, H.; Cheng, Z.; Chen, M.; He, P.; Zhou, H. Germanium thin film protected lithium aluminum germanium phosphate for solid-state Li batteries. *Adv. Energy Mater.* **2018**, *8*, 1702374. [CrossRef]
30. Wei, X.; Rechtin, J.; Olevsky, E. The fabrication of all-solid-state lithium-ion batteries via spark plasma sintering. *Metals* **2017**, *7*, 372. [CrossRef]
31. Yan, B.; Zhu, Y.; Pan, F.; Liu, J.; Lu, L. $Li_{1.5}Al_{0.5}Ge_{1.5}(PO_4)_3$ Li-ion conductor prepared by melt-quench and low temperature pressing. *Solid State Ionics* **2015**, *278*, 65–68. [CrossRef]
32. Huang, L.Z.; Wen, Z.Y.; Jin, J.; Liu, Y. Preparation and characterization of PEO-LATP/LAGP ceramic composite electrolyte membrane for lithium batteries. *J. Inorg. Mater.* **2012**, *27*, 249–252. [CrossRef]

33. Guo, Q.; Han, Y.; Wang, H.; Xiong, S.; Li, Y.; Liu, S.; Xie, K. New class of LAGP-based solid polymer composite electrolyte for efficient and safe solid-state lithium batteries. *ACS Appl. Mater. Interfaces* **2017**, *9*, 41837–41844. [CrossRef]
34. Kerman, K.; Luntz, A.; Viswanathan, V.; Chiang, Y.-M.; Chen, Z. Review—practical challenges hindering the development of solid state Li ion batteries. *J. Electrochem. Soc.* **2017**, *164*, A1731–A1744. [CrossRef]
35. Zhu, H.; Liu, J. Emerging applications of spark plasma sintering in all solid-state lithium-ion batteries and beyond. *J. Power Sources* **2018**, *391*, 10–25. [CrossRef]
36. Duan, S.; Jin, H.; Yu, J.; Esfahani, E.N.; Yang, B.; Liu, J.; Ren, Y.; Chen, Y.; Lu, L.; Tian, X.; et al. Non-equilibrium microstructure of $Li_{1.4}Al_{0.4}Ti_{1.6}(PO_4)_3$ superionic conductor by spark plasma sintering for enhanced ionic conductivity. *Nano Energy* **2018**, *51*, 19–25. [CrossRef]
37. Dumont-Botto, E.; Bourbon, C.; Patoux, S.; Rozier, P.; Dolle, M. Synthesis by spark plasma sintering: A new way to obtain electrode materials for lithium ion batteries. *J. Power Sources* **2011**, *196*, 2274–2278. [CrossRef]
38. Aboulaich, A.; Bouchet, R.; Delaizir, G.; Seznec, V.; Tortet, L.; Morcrette, M.; Rozier, P.; Tarascon, J.-M.; Viallet, V.; Dollé, M. A new approach to develop safe all-inorganic monolithic Li-ion batteries. *Adv. Energy Mater.* **2011**, *1*, 179–183. [CrossRef]
39. Pérez-Estébanez, M.; Isasi-Marín, J.; Rivera-Calzada, A.; León, C.; Nygren, M. Spark plasma versus conventional sintering in the electrical properties of NASICON-type materials. *J. Alloys Compd.* **2015**, *651*, 636–642. [CrossRef]
40. Gao, Y.-X.; Wang, X.-P.; Sun, Q.-X.; Zhuang, Z.; Fang, Q.-F. Electrical properties of garnet-like lithium ionic conductors $Li_{5+x}Sr_xLa_{3-x}Bi_2O_{12}$ fabricated by spark plasma sintering method. *Front. Mater. Sci.* **2012**, *6*, 216–224. [CrossRef]
41. Kali, R.; Mukhopadhyay, A. Spark plasma sintered/synthesized dense and nanostructured materials for solid-state Li-ion batteries: overview and perspective. *J. Power Sources* **2014**, *247*, 920–931. [CrossRef]
42. Cui, Y.; Mahmoud, M.M.; Rohde, M.; Ziebert, C.; Seifert, H.J. Thermal and ionic conductivity studies of lithium aluminum germanium phosphate solid-state electrolyte. *Solid State Ionics* **2016**, *289*, 125–132. [CrossRef]
43. Zhao, E.; Ma, F.; Guo, Y.; Jin, Y. Stable LATP/LAGP double-layer solid electrolyte prepared via a simple dry-pressing method for solid state lithium ion batteries. *RSC Adv.* **2016**, *6*, 92579–92585. [CrossRef]
44. Meesala, Y.; Chen, C.-Y.; Jena, A.; Liao, Y.-K.; Hu, S.-F.; Chang, H.; Liu, R.-S. All-solid-state Li-ion battery Using $Li_{1.5}Al_{0.5}Ge_{1.5}(PO_4)_3$ as electrolyte without polymer interfacial adhesion. *J. Phys. Chem. C* **2018**, *122*, 14383–14389. [CrossRef]
45. Zhang, Z.; Chen, S.; Yang, J.; Liu, G.; Yao, X.; Cui, P.; Xu, X. Stable cycling of all-solid-state lithium battery with surface amorphized $Li_{1.5}Al_{0.5}Ge_{1.5}(PO_4)_3$ electrolyte and lithium anode. *Electrochim. Acta* **2019**, *297*, 281–287. [CrossRef]

 *nanomaterials*

Article

One-Dimensional Porous Silicon Nanowires with Large Surface Area for Fast Charge–Discharge Lithium-Ion Batteries

Xu Chen [1], Qinsong Bi [1], Muhammad Sajjad [1], Xu Wang [2], Yang Ren [1], Xiaowei Zhou [1,*], Wen Xu [1,*] and Zhu Liu [1,3,*]

1 Department of Physics and Astronomy, Yunnan University, Kunming 650091, Yunnan, China; chenxu@ynu.edu.cn (X.C.); ggmbqs1011@gmail.com (Q.B.); sajjadfisica@gmail.com (M.S.); reny@lzu.edu.cn (Y.R.)

2 Materials Science and Engineering Program, University of Houston, Houston, TX 77204, USA; xwang65@uh.edu

3 Micro and Nano-materials and Technology Key Laboratory of Yunnan Province, Kunming 650091, Yunnan, China

* Correspondence: zhouxiaowei@ynu.edu.cn (X.Z.); wenxu_issp@aliyun.com (W.X.); zhuliu@ynu.edu.cn (Z.L.)

Received: 10 April 2018; Accepted: 25 April 2018; Published: 27 April 2018

Abstract: In this study, one-dimensional porous silicon nanowire (1D–PSiNW) arrays were fabricated by one-step metal-assisted chemical etching (MACE) to etch phosphorus-doped silicon wafers. The as-prepared mesoporous 1D–PSiNW arrays here had especially high specific surface areas of 323.47 $m^2 \cdot g^{-1}$ and were applied as anodes to achieve fast charge–discharge performance for lithium ion batteries (LIBs). The 1D–PSiNWs anodes with feature size of ~7 nm exhibited reversible specific capacity of 2061.1 $mAh \cdot g^{-1}$ after 1000 cycles at a high current density of 1.5 $A \cdot g^{-1}$. Moreover, under the ultrafast charge–discharge current rate of 16.0 $A \cdot g^{-1}$, the 1D–PSiNWs anodes still maintained 586.7 $mAh \cdot g^{-1}$ capacity even after 5000 cycles. This nanoporous 1D–PSiNW with high surface area is a potential anode candidate for the ultrafast charge–discharge in LIBs with high specific capacity and superior cycling performance.

Keywords: mesoporous structure; high surface areas; ultrafast; long-cycling life

1. Introduction

Nowadays, the latest research works have paid more attention to the increasing demand of energy storage devices in various fields, such as supercapacitors, electric vehicles, portable electronics, and lithium-ion batteries (LIBs) [1–5]. The LIBs, which possess significantly higher energy density compared to sodium-ion batteries, lead-acid batteries, and aqueous nickel-based systems, have become the dominant power storage device [6–10]. Many novel technologies have been applied to improve the cyclability and electrochemical performance of the anode materials for LIBs by designing and utilizing various nanostructures. Silicon is a promising anode for LIBs due to its highest theoretical specific capacity (4200.0 $mAh \cdot g^{-1}$) [11–13], ten times larger than that of the commercial graphite anode (372.0 $mAh \cdot g^{-1}$), with low charge–discharge potential (~0.4 V, vs. Li/Li^+) [14–19]. However, the tremendous volume change (>300%) of silicon anodes results in pulverization during fast charge lithiation and discharge delithiation processes, hindering their application as anode materials for LIBs [20–22].

Up to now, numerous strategies have been proposed to address the pulverization of the silicon anode by utilizing 0D to 2D nanostructures, such as hollow nanospheres, nanowalls [9], nanorods [23], nanosheets [24], nanotubes [16], porous silicon [25–33], and other silicon-based composites including

Si–carbon nanofibers [34], Si–C [35,36], Si–graphene [37], Cu–Si core–shell [38], and conducting polymers [39]. The nanostructured silicon as anodes for LIBs reported so far has achieved the high specific capacity of 1390~3200 mAh·g^{-1} and high specific surface area [11,12,20,40,41]. For instance, Li et al. fabricated silicon nanowires with the high surface area of 219.4 m^2·g^{-1}, which show encouraging cycling performance as an anode with reversible capacity of 2111 mAh·g^{-1} at a relatively small current density of 0.8 A·g^{-1} after 50 cycles [42]. Cui et al. first reported that LIBs using silicon nanowires (SiNWs) with diameter of around 50 nm as anodes could achieve the theoretical charge capacity (~3200.0 mAh·g^{-1}), but only maintain a discharge capacity around 75% of its original value over 10 cycles under even the small specific current of 0.2 A·g^{-1} due to pulverization [41]. Peng et al. reported that the purely electroless-etched SiNWs anodes of LIBs could achieve the large discharge capacity of 0.55 mAh·cm^{-2} over three cycles [43]. To address these issues, Bao's group increased the fast charge–discharge current of 4.0 A·g^{-1} to achieve reversible capacity of ~626.5 mAh·g^{-1} after 200 cycles, by fabricating a nanoporous silicon particle-coated carbon layer with the feature size of ~50 nm and high surface area (303.2 m^2·g^{-1}) [40]. The porous nanoparticle structure with small feature size and high surface area improved cycling ability, but with capacity degrading at fast charge–discharge current density due to the coated carbon hindering fast diffusion of Li$^+$. Hence, nanoporous silicon structures without coating and with smaller feature size could obtain high capacity with superior cycling performance at fast charge–discharge current density.

Herein, we developed a novel 1D–PSiNWs anode with high specific area and small feature size without coating, to realize fast charge–discharge at the current density of 16.0 A·g^{-1}. The schematic diagram of 1D–PSiNWs by one-step metal-assisted chemical etching (MACE) based on phosphorus-doped silicon wafers at 50 °C is shown in Figure 1 [44,45]. The formation mechanism of 1D–PSiNWs prepared by direct etching of phosphorus-doped silicon wafers is analyzed in Figure S1. It is noteworthy that the as-prepared 1D–PSiNWs have a high specific surface area of 323.47 m^2·g^{-1} and a feature size of ~7 nm through the Brunauer–Emmett–Teller (BET) method. Moreover, those with optimized pore structure anodes exhibit reversible specific capacity of 2061.1 mAh·g^{-1} at a specific current of 1.5 A·g^{-1} after 1000 cycles. Our work provides a highly efficient way for the fabrication of fast charge–discharge anode materials for LIBs.

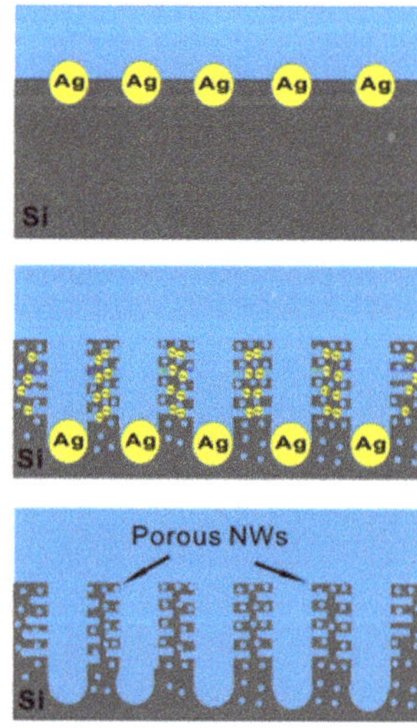

Figure 1. Schematic diagram illustrating the etching procedure of 1D–PSiNWs on silicon wafers.

2. Experimental Methods

2.1. Preparation of 1D–PSiNWs

The preparation process of 1D–PsiNWs involves the use of one-step MACE of phosphorus-doped silicon wafers. N-type silicon wafers with <100> oriented (0.001–0.005 $\Omega \cdot$ cm) were cut into pieces with measurements of 2.0 × 2.0 cm^2 (Lijing Silicon Materials Co., Ltd., Quzhou, China). The fabrication process of 1D–PSiNWs was as follows: (1) the silicon wafer pieces were cleaned by ultrasonication in acetone (10 min), ethanol (10 min), and deionized water several times (DI water, 18.25 M $\Omega \cdot$ cm), respectively. Then, silicon wafer pieces were dipped in H_2SO_4/H_2O_2 solution (volume ratio of 97% H_2SO_4/30% H_2O_2 = 3:1) at 96 °C (1 min) and completely washed with DI water. (2) Cleaned silicon wafer pieces were immersed in the mixture of 4.6 M HF solution and 0.02 M $AgNO_3$ in sealed vessels and treated for 2 h at 50 °C in dark condition. Immediately, the thick dendritic silver film was coated on etched silicon substrates. (3) The etched silicon substrates were totally dipped into HNO_3 (volume ratio of HNO_3/DI water = 1:1) solution for 4 h to remove silver dendritic layer, then transferred into HF (5 wt %) solution to control 1D–PSiNW length and for a suitable time to ensure that the newly formed SiO_2 was removed. The perpendicularly ordered 1D–PSiNWs could be discovered on silicon substrates after flaking off the dendritic silver film. (4) The resulting substrates were thoroughly washed with DI water and ethanol, and vacuum-dried at 70 °C for 8 h.

2.2. Material Characterization

The as-synthesized samples were characterized by an X-ray diffractometer (XRD, Rigaku TTRIII, Rigaku Corporation, Tokyo, Japan) using Cu-Kα radiation (1.5406 Å, 35 kV). The morphology and structure of the 1D–PSiNWs were investigated by scanning electron microscopy (SEM, FEI QUANTA200, FEI NanoPorts, Miami, FL, USA) coupled with energy dispersive X-ray spectroscopy (EDX), field emission scanning electron microscopy (FESEM, FEI Nova Nano SEM 450, Rigaku Corporation, Tokyo, Japan), and transmission electron microscopy (TEM, JEM-2100, Rigaku Corporation, Tokyo, Japan). Samples of TEM were obtained using ultrasonically oscillating etched silicon wafers in ethanol solution. The nitrogen adsorption and desorption isotherms were obtained using the Brunauer–Emmett–Teller (BET) method at 77 K after degassing the samples at 200 °C for 4 h by an analyzer (Quantachrome, Quadrasorb evo, Boynton Beach, FL, USA). The Barrett–Joyner–Halenda (BJH) method was applied to adsorption branches of isotherms to obtain pore diameter distributions. Chemical bonding analysis of the 1D–PSiNW surface layer was carried out using X-ray photoelectron spectroscopy (XPS, Thermo Scientific K-ALPHA$^+$, Thermo Fisher Scientific, Waltham, MA, USA).

2.3. Electrochemical Characterization

The working electrodes were made up by mixing the active material (1D–PSiNWs), conductive agent (carbon black), and binder (sodium alginate) in a weight ratio of 50:30:20. The loading density of 1D–PSiNW active materials was $\approx$ 0.15 mg $\cdot$ cm^{-2}. Coin-type cells (CR2025) were assembled in an argon-filled glove box (MIKROUNA super, $O_2 \leq 0.1$ ppm, $H_2O \leq 0.1$ ppm) while applying lithium metal as the counter electrode, a polypropylene (PP) microporous film as the separator, and $LiPF_6$ (1 M) in ethylene carbonate (EC)-dimethyl carbonate (DMC) and diethyl carbonate (DEC) (volume ratio of EC:DMC:DEC = 1:1:1) as the electrolyte. The galvanostatic measurements were carried out on a LAND-CT2001A battery tester with a voltage window of 0.01–2.0 V at various current rates. The cyclic voltammetry was performed using an electrochemical workstation (chi604e) between 0.01 and 2.0 V at a scan rate of 0.1 mV $\cdot$ s^{-1}. The electrochemical impedance spectra (EIS) of the cells were measured on the electrochemical workstation (chi604e) at the frequency range of 0.1 Hz–100 kHz. Nyquist plots derived from EIS were simulated by using Z-view software.

3. Result and Discussion

Figure 2 shows the XRD patterns of the 1D–PSiNWs. The main observed peaks shown in Figure 2 can be indexed to the face-centered-cubic (fcc) structure of silicon (JCPD Card No. 27-1402), which belongs to the space group Fd-3m (No. 227). The peak of 38.17° indicates the existence of a tiny minority of Ag in the obtained 1D–PSiNWs after HNO_3 treatment. The morphologies and EDX analysis of 1D–PSiNWs coated by dendritic Ag film after etching were obtained by scanning electron microscopy (SEM) as presented in Figure S2a.

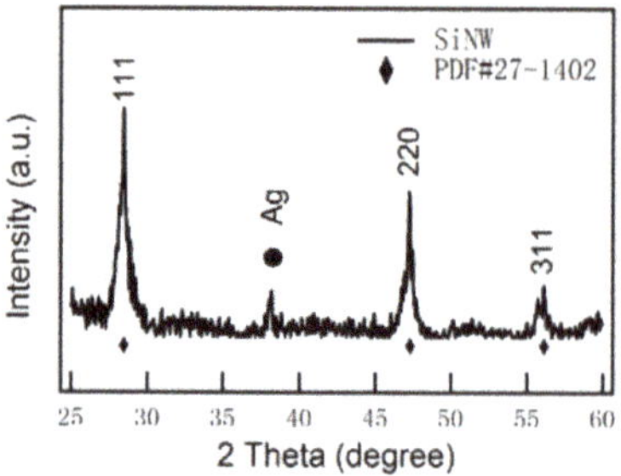

Figure 2. X-ray diffraction patterns of 1D–PSiNWs.

High-quality 1D–PSiNW arrays can be produced on the n-type Si wafer by simply immersing the wafer into $HF/AgNO_3$ solution for an appropriate etching time and temperature. Figure 3 shows FESEM (Figure 3a,b) and TEM (Figure 3c,f) images of 1D–PSiNWs. The cross section profile in Figure 3a confirms that the 1D–PSiNW arrays are uniform on the entire wafer surface. The enlarge cross-section in Figure 3b indicates that the diameters of the vertically well-aligned arrays are in the range of 60.0–500.0 nm. Mesoporous holes are uniformly distributed on the surface of each nanowire. From the magnified cross-section image shown in Figure S2b, some orderly congregated bundles composed of 1D–PSiNWs can be clearly seen. The lengths of the as-prepared 1D–PSiNWs still near 7.0 μm after HF treatment. Figure S3c is an enlarged top view of 1D–PSiNWs, which indicates a uniform surface. The TEM images in Figure 3c–f demonstrate that the nanowires are highly porous at the surface, with both pore diameter and wall thickness around 7 nm. The polycrystalline structure was confirmed by the circular diffraction pattern in the selected area electron diffraction (SAED) taken on a single porous nanowire as illustrated in Figure 3c. Some parts of 1D–PSiNWs also have monocrystal properties, displaying clear lattice fringes corresponding to a silicon (111) lattice with an interplanar crystal spacing of 3.14 Å, as presented in Figure 3f. The phosphorus dopants provide electrons which facilitate the etching process and leave holes on the silicon nanowire surface, resulting in the coexistence of crystalline and amorphous states. Figure S1 illustrates a simplified quantitative model for the detailed growth mechanism of 1D–PSiNWs in aqueous $HF/AgNO_3$ solution. The Fermi level of N-type silicon is more positive than the redox potential of Ag/Ag^+, leading to majority carrier electrons which will transfer to the silicon/solution interface. Thus, the holes from the oxidant will be injected into the valence band of silicon with the Ag deposition or reduction of H^+ which induces silicon substrate oxidization and dissolution, leading to SiNW growth. The etching process can be described as two main simultaneous electrochemical reactions [46,47]:

$$4Ag^+ + 4e^- \rightarrow 4Ag$$

$$Si + 6F^- \rightarrow [SiF_6]^{2-} + 4e^-$$

Surface area measurements are typically based on N_2 sorption at 77.3 K, and the Brunauer–Emmer–Teller (BET) model is typically used to interpret the data. The N_2 sorption isotherm and the pore diameter size distribution of 1D–PSiNWs are shown in Figure 4. Capillary condensation occurs in the higher P/P_0 region ($P/P_0 > 0.15$), and hysteresis can be observed, which conforms to the typical isothermal curve of mesopores (type IV). A pore size distribution analysis by Barrett–Joyner–Halenda (BJH) methods showed that there is a pore diameter distribution of 1D–PSiNWs which is mesoporous at the range of 3.8–80.5 nm. The maximum probability for feature size distribution is ~7 nm in diameter (inset). This result is in a good agreement with the TEM observation. The 1D–PSiNWs exhibit exceptionally high specific surface area of 323.47 $m^2 \cdot g^{-1}$. This value is, significantly, the same as the maximum value reported for pure porous silicon materials by metal-assisted chemical etching (MACE) [33].

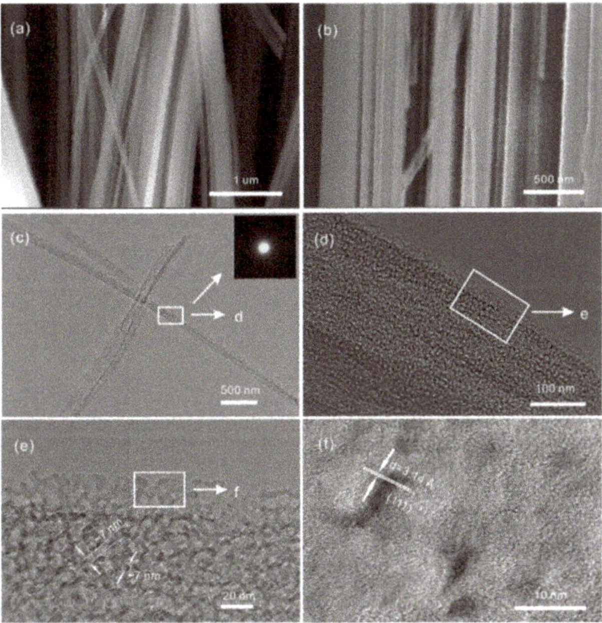

Figure 3. SEM (**a**,**b**), TEM (**c**,**d**), and HRTEM (**e**,**f**) images of 1D–PSiNW arrays etched.

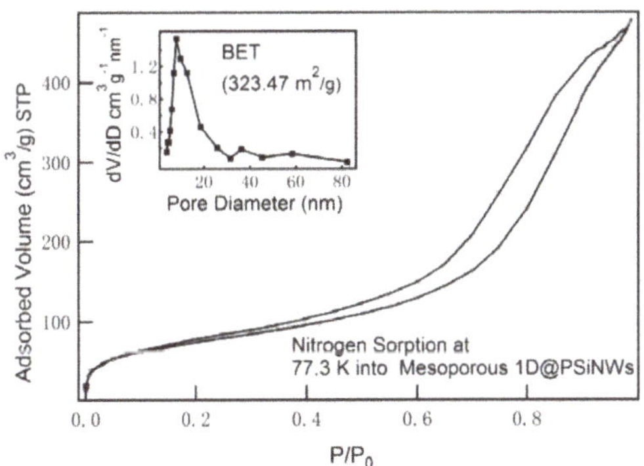

Figure 4. Nitrogen adsorption isotherm at 77.3 K (line, adsorption; triangular, desorption) and pore size distribution curve (inset) of 1D–PSiNWs depicted after fitting by the Barrett–Joyner–Halenda (BJH) model.

The whole XPS spectra of 1D–PSiNWs, which are calibrated with the C1s line of adventitious carbon at 284.8 eV, are presented in Figure S3 with the four main components fitted (Si at 99.4, SiO_2 at 103.6, O_2 at 532.9, and $[SiF_6]^{2-}$ at 687.1 eV) [22]. Figure 5a presents the Si_{2p}-amplified high-resolution XPS spectrum, which reveals two peaks centered at 99.4 eV and 103.6 eV that can be attributed to silicon and SiO_2 components, respectively. The typical $Si_{2p1/2}$ and $Si_{2p3/2}$ spectra are overlapped partly, giving an asymmetric peak shape at 99.4 eV. Figure 5b shows the O1s high-resolution XPS spectrum with a narrow signal peak at 532.9 eV, indicating the coexistence of unique oxygen chemical environments on the surface of the structure. The binding energy of the O1s signal at 532.9 eV is in full compliance with the chemical state of SiO_2, which can be assigned to the Si–O bond. A weak additional peak corresponding to $[SiF_6]^{2-}$ at 687.1 eV is observed in Figure 5c. Actually, small quantities of organic fluorine which are not totally eliminated during the etching process are often found. In fact, as observed in the XPS spectra, the presence of oxygen atoms and fluorine atoms is due to the partial oxidation of 1D–PSiNWs surface and the formation of Si–O bonds/$[SiF_6]^{2-}$ during the etching process of the silicon wafer on the 1D–PSiNW surface. The XPS survey scans did not detect any other impurities except oxygen and fluorine. This also shows that the silver elements reflected by the XRD pattern do not exist on the surface of the 1D–PSiNWs, but inside the deep holes.

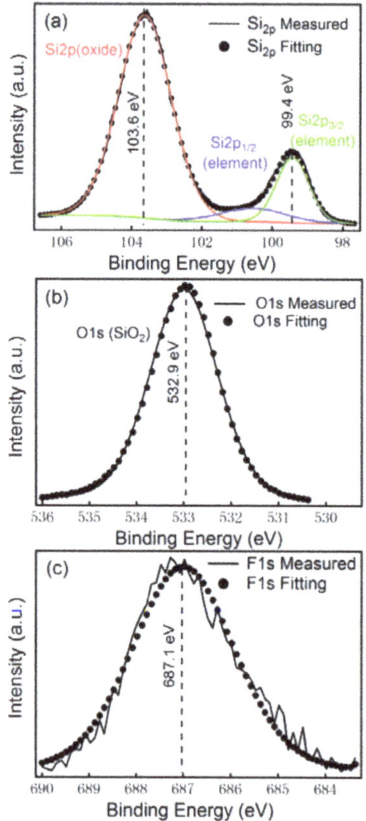

Figure 5. X-ray photoelectron spectroscopy (XPS) survey of 1D–PSiNWs using (**a**) Si_{2p}, (**b**) O_{1s}, and (**c**) F_{1s}.

The electrochemical performances of 1D–PSiNWs were investigated by repeated charge–discharge cycling at room temperature in a coin-type half-cell using Li metal as the anode. The electrochemical reactions occurring in a coin-type half-cell is described by the following formula [32,48]:

$$\text{Si electrode}: \ xLi^{+} + Si + xe^{-} \leftrightarrow Li_xSi, \ 0 \leq x \leq 4.4$$

$$\text{Li electrode}: \ Li \leftrightarrow Li^{+} + e^{-}$$

Cyclic voltammetry (CV) curves at the first and second charge–discharge cycles of anodes based on 1D–PSiNWs were acquired in the potential window from 0.01 V to 2.0 V (vs Li/Li^{+}) at a scan rate of 0.1 $mV{\cdot}s^{-1}$ (Figure 6). The voltage profile observed was consistent with previous silicon anode studies, with a long flat plateau during the first charge, during which crystalline silicon reacted with Li^{+} to form amorphous Li_xSi [12,40,41]. During the first discharge cycle, the peak at 0.8 V, which is absent at the second cycle, leads to the formation of an extremely large solid/electrolyte interphase (SEI) on the surface of the porous silicon nanowire electrode, which leads to irreversible capacity loss. We can see the peak at 0.16 V during the first discharge (Li alloy), which is due to the phase transition of silicon to the amorphous lithium-rich $Li_{15}Si_4$ structure. During the first charge process (Li dealloy), two broad peaks were observed at 0.28 V and 0.47 V, which can be attributed to the phase transition between amorphous Li_xSi and amorphous silicon. Upon the second discharge and charge, peaks were also observed at 0.18 V and 0.28 V/0.47 V, which are the same as that of the first discharge, respectively. The CVs from the 200th to 203rd cycles are shown in Figure S4. The charge–discharge curves from the 200th to 203rd cycles almost overlap each other, thus indicating highly stable electrochemical cycling performance.

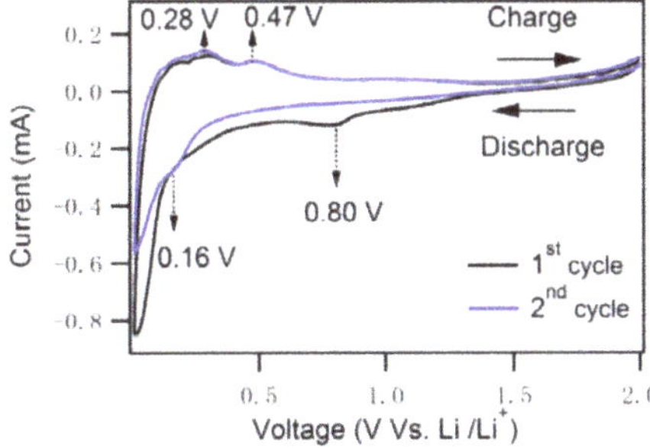

Figure 6. Cyclic voltammograms of 1D–PSiNWs anodes for the first and second cycle at scan rate of 0.1 $mV{\cdot}s^{-1}$ (voltage range: 0.01 V–2.0 V).

Representative charge–discharge profiles of the electrode based on 1D–PSiNWs with 100 cycles at a rate of 1 $A{\cdot}g^{-1}$ in the potential range of 0.01–2.0 V are shown in Figure 7a. The electrode based on 1D–PSiNWs exhibits a discharge capacity of 4487.2, 2592.0, 2350.3, and 2004.5 $mAh{\cdot}g^{-1}$ at the first, 10th, 50th, and 100th cycle, respectively. The first discharge and charge capacities of the as-prepared 1D–PSiNWs anodes are 4487.2 $mAh{\cdot}g^{-1}$ and 2534.5 $mAh{\cdot}g^{-1}$, respectively; thus indicating an initial coulombic efficiency of 56.5%. The irreversible capacity is due to the solid/electrolyte interphase (SEI) film forming on all surfaces of the 1D–PSiNWs with high specific surface area of 323.47 $m^2{\cdot}g^{-1}$. Therefore, the first discharge irreversible capacity loss of ~1953.0 $mAh{\cdot}g^{-1}$ could mainly originate from the reduction of electrolyte and the formation of a SEI on the surface of an electrode, and from the irreversible insertion of lithium ions into silicon nanowires [42].

To show the advantage and stability of 1D–PSiNWs, the cycling performance is shown in Figure 7b. The 1D–PSiNW electrode exhibited excellent cycling performance with a reversible capacity of 2061.1 $mAh{\cdot}g^{-1}$ over 1000 cycles under a high current density of 1.5 $A{\cdot}g^{-1}$, with 99.7% coulombic efficiency. The rate capability (0.4–4.0 A·g^{-1}) of the 1D–PSiNW anodes was evaluated in the voltage range of 0.01–2.0 V, as presented in Figure 7c. The specific discharge capacities of the 1D–PSiNWs for the last cycles are 2706.5, 2314.0, 2101.2, and 1667.0 mAh·g^{-1} at current densities of 0.4, 1.0, 2.0, and 4.0 A·g^{-1}, respectively. After 60 charge–discharge cycles under a high current density of 4.0 A·g^{-1}, a specific discharge capacity as high as 1667.0 mAh·g^{-1} was maintained. Increasing the current density to 8.0 A·g^{-1}, a discharge capacity of 1650.6 mAh·g^{-1} after 60 cycles was delivered (Figure S5). After going through super-long 5000 cycles at 16.0 A·g^{-1}, as shown in Figure 7d, a discharge capacity of 586.7 mAh·g^{-1} was still retained, which was much larger compared with that of graphite (~372.0 mAh·g^{-1}). The slight increase in the capacity for 1D–PSiNWs in the initial 100 cycles is associated with the gradual activation of the silicon host (the same phenomenon can also be seen in Figure 7b). It is reported that there is a slight increase in the capacity until full lithiation of the active silicon [46,48–50]. The rapid fading in capacity for the silicon powder or nanowire was due to the large volume change during lithium alloying and dealloying processes, leading to fragmentation and an electrical disconnection between particles. The improved performance of pure 1D–PSiNWs can be attributed to the interconnected porous network structure, which offers sufficient void space to accommodate the large volume change. In addition, a huge quantity of mesopores inside silicon nanowires act as smooth channels for the fast Li^+/electron transfer, and are conducive to the rate capability and cycling stability of 1D–PSiNW anodes.

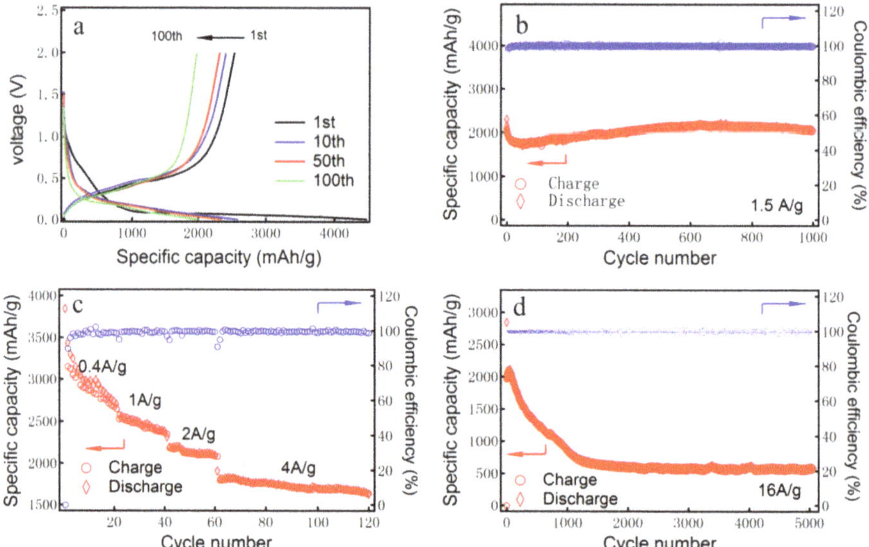

Figure 7. Results of electrochemical performance for 1D–PSiNW anodes. (**a**) Galvanostatic charge/discharge profiles between 0.01 V and 2.0 V vs. Li/Li^+ for the first, 10th, 50th, and 100th cycles at a current density of 1.0 A·g^{-1}. (**b**) Cycling performance of the as-prepared 1D–PSiNW anodes at a current density of 1.5 A·g^{-1}. (**c**) Rate performance of 1D–PSiNW anodes at the current density of 0.4/1.0/2.0/4.0 A·g^{-1}. (**d**) Electrochemical performance of 1D–PSiNW anodes with 5000 cycles at a current density of 16.0 A·g^{-1}.

The electrochemical impedance (EIS) of the 1D–PSiNW electrode is investigated in Figure 8 to gain further insights into the improved cycling performance. The Nyquist plots obtained from open-circuit voltage (E = 2.0 V) manifested that the charge transfer resistance of the electrode is minimal for 1D–PSiNW anodes. The fitting equivalent circuit for the cell system is depicted in the inset of Figure 8. In the equivalent circuit for the fresh cell before cycling, shown in Figure 8a, R_s (7.7 Ω) is the bulk resistance of the 1D–PSiNW electrode, electrolyte, and separator, corresponding to the intercept value of the semicircle with the real axis at the high-frequency region. CPE is the constant phase element, and R_{ct} (66 Ω) is the charge transfer resistance, corresponding to the diameter of the semicircle at high frequency. W is the Warburg impedance arising from the semi-infinite diffusion of Li^+ ions in the bulk of the electrode, which is generally indicated by a sloping straight line at the low-frequency region [51,52]. As shown in Figure 8b, the bulk resistance (R_s) and the charge transfer resistance (R_{ct}) are 6.7 Ω and 104 Ω for the 1D–PSiNW electrode after 200 cycles at a current density of 8.0 A·g^{-1}, respectively. There is no significant resistance change in electrode conductivity after 200 cycles, indicating that the 1D–PSiNWs possess excellent conductivity retention. The 1D–PSiNW electrodes have excellent electron/ion conductivity.

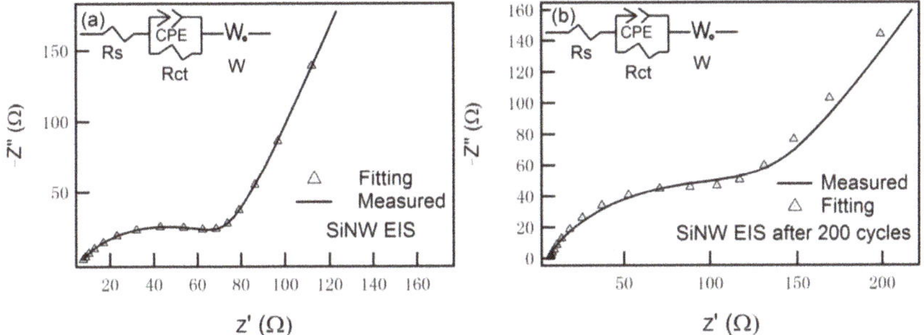

Figure 8. Typical electrochemical impedance spectra of 1D–PSiNW anodes measured at open-circuit voltage E ≈ 2.0 V (Li/Li^+): (**a**) fresh cell; (**b**) after 200 cycles at a current density of 8.0 A·g^{-1} (inset is the equivalent circuit used to fit the electrochemical impedance (EIS)).

4. Conclusions

In conclusion, we fabricated novel 1D–PSiNW anodes with the high specific area of 323.47 m^2·g^{-1} by a one-step silver-assisted chemical etching method. The TEM image demonstrated that the nanowires were several μm length and 60–500 nm in diameter with highly uniform porosity at the surface, with both pore diameter and wall thickness around 7 nm. This small pore size of the silicon nanoporous structure offered sufficient void space to accommodate the large volume change. Our designed 1D–PSiNWs as anodes for LIBs show a reversible specific capacity of 2061.1 mAh·g^{-1} after 1000 cycles under a fast charge–discharge condition at 1.5 A·g^{-1}. Even after 5000 cycles, a reversible capacity of 586.7 mAh·g^{-1} was retained at the ultrafast charge–discharge current density of 16.0 A·g^{-1}. The small feature size acted as a short Li^+/electron conductive path during the lithiation/delithation processes. The superior electrochemical performance and excellent cycling life of the nanoporous 1D–PSiNW anodes were attributed to the 1D structure with uniform interconnected nanoporous channels existing inside the silicon nanowires.

Supplementary Materials: The following are available online at http://www.mdpi.com/2079-4991/8/5/285/s1. Figure S1: Energy-band diagrams of N-type silicon in aqueous HF/$AgNO_3$ solution. Figure S2: (a) SEM and EDX of dendritic Ag-coated 1D–PSiNWs after etching, (b) SEM cross-section of 1D–PSiNWs, and (c) top view SEM image of 1D–PSiNWs. Figure S3: XPS survey spectra of 1D–PSiNWs. Figure S4: Cyclic voltammetry curves of 1D–PSiNW anodes of 200th, 201st, 202nd cycles in the voltage window from 0.01 V to 2.0 V at rate of 0.1 mV·s^{-1}. Figure S5: The results of cycling performance of 1D–PSiNW anodes tested at a current density of 8.0 A·g^{-1}.

Author Contributions: Xu Chen designed the experiments and wrote the paper. Qinsong Bi and Muhammad Sajjad prepared and characterized the samples. Xu Wang and Yang Ren provided useful discussion. Xiaowei Zhou contributed analysis experimental data. Wen Xu and Zhu Liu conceived the study and provided critical advice for each designed experiment.

Acknowledgments: We gratefully acknowledge the financial support from the Natural Science Foundation of China (grant nos. 61664009 and 51771169) and the Youth Project of Applied Basic Research of the Yunnan Science and Technology Department (grant no. 2015FD001). This work is also funded in part by the High-End Scientific and Technological Talents Introduction Project of Yunnan Province (grant No. 2013HA019). Moreover, the authors wish to acknowledge Kuiqing Peng for his help in useful discussions about reaction principle. Gang Cao provided useful discussion. Long Li and Ying Huang designed the experiments and battery assembly test.

Conflicts of Interest: The authors declare no conflicts of interest.

References

1. Armand, M.; Tarascon, J.-M. Building better batteries. *Nature* **2008**, *451*, 652–657. [CrossRef] [PubMed]
2. Dunn, B.; Kamath, H.; Tarascon, J.-M. Electrical energy storage for the grid: A battery of choices. *Science* **2011**, *334*, 928–935. [CrossRef] [PubMed]
3. Simon, P.; Gogotsi, Y. Materials for electrochemical capacitors. *Nat. Mater.* **2008**, *7*, 845–854. [CrossRef] [PubMed]
4. Repp, S.; Harputlu, E.; Gurgen, S.; Castellano, M.; Kremer, N.; Pompe, N.; Worner, J.; Hoffmann, A.; Thomann, R.; Emen, F.M.; et al. Synergetic effects of Fe^{3+} doped spinel $Li_4Ti_5O_{12}$ nanoparticles on reduced graphene oxide for high surface electrode hybrid supercapacitors. *Nanoscale* **2018**, *10*, 1877–1884. [CrossRef] [PubMed]
5. Genc, R.; Alas, M.O.; Harputlu, E.; Repp, S.; Kremer, N.; Castellano, M.; Colak, S.G.; Ocakoglu, K.; Erdem, E. High-Capacitance Hybrid Supercapacitor Based on Multi-Colored Fluorescent Carbon-Dots. *Sci. Rep.* **2017**, *7*, 11222. [CrossRef] [PubMed]
6. Zhou, X.; Chen, X.; He, T.; Bi, Q.; Sun, L.; Liu, Z. Fabrication of polypyrrole/vanadium oxide nanotube composite with enhanced electrochemical performance as cathode in rechargeable batteries. *Appl. Surf. Sci.* **2017**, *405*, 146–151. [CrossRef]
7. Guo, S.; Yi, J.; Sun, Y.; Zhou, H. Recent advances in titanium-based electrode materials for stationary sodium-ion batteries. *Energy Environ. Sci.* **2016**, *9*, 2978–3006. [CrossRef]
8. Maeda, K.; Arakawa, N.; Fukui, K.; Kuramochi, H. Fast charging of lead–acid batteries enabled by high-pressure crystallization. *PCCP* **2014**, *16*, 4911–4916. [CrossRef] [PubMed]
9. Gong, Y.; Zhang, P.; Jiang, P.; Lin, J. Metal (II) complex based on 5-[4-(1*H*-imidazol-1-yl) phenyl]-2*H*-tetrazole: The effect of the ligand on the electrodes in a lead-acid battery. *CrystEngComm* **2015**, *17*, 1056–1064. [CrossRef]
10. Zhou, X.; Chen, X.; Bi, Q.; Luo, X.; Sun, L.; Liu, Z. A scalable strategy to synthesize TiO_2-V_2O_5 nanorods as high performance cathode for lithium ion batteries from VOx quasi-aerogel and tetrabutyl titanate. *Ceram. Int.* **2017**, *43*, 12689–12697. [CrossRef]
11. Jin, Y.; Tan, Y.; Hu, X.; Zhu, B.; Zheng, Q.; Zhang, Z.; Zhu, G.; Yu, Q.; Jin, Z.; Zhu, J. Scalable Production of the Silicon–Tin Yin-Yang Hybrid Structure with Graphene Coating for High Performance Lithium-Ion Battery Anodes. *ACS Appl. Mater. Interfaces* **2017**, *9*, 15388–15393. [CrossRef] [PubMed]
12. Zuo, X.; Zhu, J.; Müller-Buschbaum, P.; Cheng, Y.-J. Silicon based lithium-ion battery anodes: A chronicle perspective review. *Nano Energy* **2017**, *31* (Suppl. C), 113–143. [CrossRef]
13. Schmerling, M.; Fenske, D.; Peters, F.; Schwenzel, J.; Busse, M. Lithiation Behavior of Silicon Nanowire Anodes for Lithium-Ion Batteries: Impact of Functionalization and Porosity. *ChemPhysChem* **2017**, *19*, 123–129. [CrossRef] [PubMed]
14. Chen, X.; Gerasopoulos, K.; Guo, J.; Brown, A.; Wang, C.; Ghodssi, R.; Culver, J.N. Virus-Enabled Silicon Anode for Lithium-Ion Batteries. *ACS Nano* **2010**, *4*, 5366–5372. [CrossRef] [PubMed]
15. Zhang, H.; Braun, P.V. Three-Dimensional Metal Scaffold Supported Bicontinuous Silicon Battery Anodes. *Nano Lett.* **2012**, *12*, 2778–2783. [CrossRef] [PubMed]
16. Wen, Z.; Lu, G.; Mao, S.; Kim, H.; Cui, S.; Yu, K.; Huang, X.; Hurley, P.T.; Mao, O.; Chen, J. Silicon nanotube anode for lithium-ion batteries. *Electrochem. Commun.* **2013**, *29* (Suppl. C), 67–70. [CrossRef]
17. Qin, J.; Wu, M.; Feng, T.; Chen, C.; Tu, C.; Li, X.; Duan, C.; Xia, D.; Wang, D. High rate capability and long cycling life of graphene-coated silicon composite anodes for lithium ion batteries. *Electrochim. Acta* **2017**, *256* (Suppl. C), 259–266. [CrossRef]
18. Zhang, L.; Hu, X.; Chen, C.; Guo, H.; Liu, X.; Xu, G.; Zhong, H.; Cheng, S.; Wu, P.; Meng, J.; et al. In operando mechanism analysis on nanocrystalline silicon anode material for reversible and ultrafast sodium storage. *Adv. Mater.* **2017**, *29*. [CrossRef] [PubMed]

19. Piwko, M.; Kuntze, T.; Winkler, S.; Straach, S.; Härtel, P.; Althues, H.; Kaskel, S. Hierarchical columnar silicon anode structures for high energy density lithium sulfur batteries. *J. Power Sources* **2017**, *351*, 183–191. [CrossRef]
20. Zhu, J.; Yang, J.; Xu, Z.; Wang, J.; Nuli, Y.; Zhuang, X.; Feng, X. Silicon anodes protected by a nitrogen-doped porous carbon shell for high-performance lithium-ion batteries. *Nanoscale* **2017**, *9*, 8871–8878. [CrossRef] [PubMed]
21. Ge, M.Y.; Rong, J.P.; Fang, X.; Zhou, C.W. Porous Doped Silicon Nanowires for Lithium Ion Battery Anode with Long Cycle Life. *Nano Lett.* **2012**, *12*, 2318–2323. [CrossRef] [PubMed]
22. Leisner, M.; Cojocaru, A.; Ossei-Wusu, E.; Carstensen, J.; Föll, H. New Applications of Electrochemically Produced Porous Semiconductors and Nanowire Arrays. *Nanoscale Res. Lett.* **2010**, *5*, 1502. [CrossRef] [PubMed]
23. Teki, R.; Datta, M.K.; Krishnan, R.; Parker, T.C.; Lu, T.-M.; Kumta, P.N.; Koratkar, N. Nanostructured Silicon Anodes for Lithium Ion Rechargeable Batteries. *Small* **2009**, *5*, 2236–2242. [CrossRef] [PubMed]
24. Okamoto, H.; Kumai, Y.; Sugiyama, Y.; Mitsuoka, T.; Nakanishi, K.; Ohta, T.; Nozaki, H.; Yamaguchi, S.; Shirai, S.; Nakano, H. Silicon nanosheets and their self-assembled regular stacking structure. *J. Am. Chem. Soc.* **2010**, *132*, 2710–2718. [CrossRef] [PubMed]
25. Wu, H.; Du, N.; Shi, X.; Yang, D. Rational design of three-dimensional macroporous silicon as high performance Li-ion battery anodes with long cycle life. *J. Power Sources* **2016**, *331*, 76–81. [CrossRef]
26. Xiao, C.; Du, N.; Shi, X.; Zhang, H.; Yang, D. Large-scale synthesis of Si@C three-dimensional porous structures as high-performance anode materials for lithium-ion batteries. *J. Mater. Chem. A* **2014**, *2*, 20494–20499. [CrossRef]
27. Kim, H.; Han, B.; Choo, J.; Cho, J. Three-Dimensional Porous Silicon Particles for Use in High-Performance Lithium Secondary Batteries. *Angew. Chem. Int. Ed.* **2008**, *47*, 10151–10154. [CrossRef] [PubMed]
28. Liu, Y.; Chen, B.; Cao, F.; Chan, H.L.W.; Zhao, X.; Yuan, J. One-pot synthesis of three-dimensional silver-embedded porous silicon micronparticles for lithium-ion batteries. *J. Mater. Chem.* **2011**, *21*, 17083–17086. [CrossRef]
29. Du, F.-H.; Wang, K.-X.; Fu, W.; Gao, P.-F.; Wang, J.-F.; Yang, J.; Chen, J.-S. A graphene-wrapped silver-porous silicon composite with enhanced electrochemical performance for lithium-ion batteries. *J. Mater. Chem. A* **2013**, *1*, 13648–13654. [CrossRef]
30. McSweeney, W.; Geaney, H.; O'Dwyer, C. Metal-assisted chemical etching of silicon and the behavior of nanoscale silicon materials as Li-ion battery anodes. *Nano Res.* **2015**, *8*, 1395–1442. [CrossRef]
31. Hwang, C.; Lee, K.; Um, H.-D.; Lee, Y.; Seo, K.; Song, H.-K. Conductive and Porous Silicon Nanowire Anodes for Lithium Ion Batteries. *J. Electrochem. Soc.* **2017**, *164*, A1564–A1568. [CrossRef]
32. Ge, M.; Rong, J.; Fang, X.; Zhang, A.; Lu, Y.; Zhou, C. Scalable preparation of porous silicon nanoparticles and their application for lithium-ion battery anodes. *Nano Res.* **2013**, *6*, 174–181. [CrossRef]
33. Hochbaum, A.I.; Gargas, D.; Hwang, Y.J.; Yang, P. Single Crystalline Mesoporous Silicon Nanowires. *Nano Lett.* **2009**, *9*, 3550–3554. [CrossRef] [PubMed]
34. Kong, J.; Yee, W.A.; Wei, Y.; Yang, L.; Ang, J.M.; Phua, S.L.; Wong, S.Y.; Zhou, R.; Dong, Y.; Li, X. Silicon nanoparticles encapsulated in hollow graphitized carbon nanofibers for lithium ion battery anodes. *Nanoscale* **2013**, *5*, 2967–2973. [CrossRef] [PubMed]
35. Guo, J.; Yang, Z.; Archer, L.A. Mesoporous silicon@ carbon composites via nanoparticle-seeded dispersion polymerization and their application as lithium-ion battery anode materials. *J. Mater. Chem. A* **2013**, *1*, 5709–5714. [CrossRef]
36. Fu, K.; Yildiz, O.; Bhanushali, H.; Wang, Y.; Stano, K.; Xue, L.; Zhang, X.; Bradford, P.D. Aligned carbon nanotube-silicon sheets: A novel nano-architecture for flexible lithium ion battery electrodes. *Adv. Mater.* **2013**, *25*, 5109–5114. [CrossRef] [PubMed]
37. Lee, J.K.; Smith, K.B.; Hayner, C.M.; Kung, H.H. Silicon nanoparticles–graphene paper composites for Li ion battery anodes. *Chem. Commun.* **2010**, *46*, 2025–2027. [CrossRef] [PubMed]
38. Qu, J.; Li, H.; Henry, J.J.; Martha, S.K.; Dudney, N.J.; Xu, H.; Chi, M.; Lance, M.J.; Mahurin, S.M.; Besmann, T.M. Self-aligned Cu–Si core–shell nanowire array as a high-performance anode for Li-ion batteries. *J. Power Sources* **2012**, *198*, 312–317. [CrossRef]

39. Wu, H.; Yu, G.; Pan, L.; Liu, N.; McDowell, M.T.; Bao, Z.; Cui, Y. Stable Li-ion battery anodes by in-situ polymerization of conducting hydrogel to conformally coat silicon nanoparticles. *Nat. Commun.* **2013**, *4*, 1943. [CrossRef] [PubMed]
40. Gao, P.; Tang, H.; Xing, A.; Bao, Z. Porous silicon from the magnesiothermic reaction as a high-performance anode material for lithium ion battery applications. *Electrochim. Acta* **2017**, *228*, 545–552. [CrossRef]
41. Chan, C.K.; Peng, H.; Liu, G.; McIlwrath, K.; Zhang, X.F.; Huggins, R.A.; Cui, Y. High-performance lithium battery anodes using silicon nanowires. *Nat. Nanotechnol.* **2008**, *3*, 31–35. [CrossRef] [PubMed]
42. Li, X.; Yan, C.; Wang, J.; Graff, A.; Schweizer Stefan, L.; Sprafke, A.; Schmidt Oliver, G.; Wehrspohn Ralf, B. Stable Silicon Anodes for Lithium-Ion Batteries Using Mesoporous Metallurgical Silicon. *Adv. Energy Mater.* **2014**, *5*, 1401556. [CrossRef]
43. Peng, K.; Jie, J.; Zhang, W.; Lee, S.-T. Silicon nanowires for rechargeable lithium-ion battery anodes. *Appl. Phys. Lett.* **2008**, *93*, 033105. [CrossRef]
44. Peng, K.Q.; Wu, Y.; Fang, H.; Zhong, X.Y.; Xu, Y.; Zhu, J. Uniform, axial-orientation alignment of one-dimensional single-crystal silicon nanostructure arrays. *Angew. Chem. Int. Ed.* **2005**, *44*, 2737–2742. [CrossRef] [PubMed]
45. Peng, K.; Fang, H.; Hu, J.; Wu, Y.; Zhu, J.; Yan, Y.; Lee, S. Metal-Particle-Induced, Highly Localized Site-Specific Etching of Si and Formation of Single-Crystalline Si Nanowires in Aqueous Fluoride Solution. *Chem. A Eur. J.* **2006**, *12*, 7942–7947. [CrossRef] [PubMed]
46. Chartier, C.; Bastide, S.; Lévy-Clément, C. Metal-assisted chemical etching of silicon in HF–H_2O_2. *Electrochim. Acta* **2008**, *53*, 5509–5516. [CrossRef]
47. Kolasinski, K.W.; Barclay, W.B.; Sun, Y.; Aindow, M. The stoichiometry of metal assisted etching (MAE) of Si in V_2O_5 + HF and HOOH + HF solutions. *Electrochim. Acta* **2015**, *158*, 219–228. [CrossRef]
48. Gao, B.; Sinha, S.; Fleming, L.; Zhou, O. Alloy Formation in Nanostructured Silicon. *Adv. Mater.* **2001**, *13*, 816–819. [CrossRef]
49. Hwa, Y.; Kim, W. S.; Yu, B. C.; Kim, J. H.; Hong, S.-H.; Sohn, H.-J. Facile synthesis of Si nanoparticles using magnesium silicide reduction and its carbon composite as a high-performance anode for Li ion batteries. *J. Power Sources* **2014**, *252*, 144–149. [CrossRef]
50. McDowell, M.T.; Ryu, I.; Lee, S.W.; Wang, C.; Nix, W.D.; Cui, Y. Studying the kinetics of crystalline silicon nanoparticle lithiation with in situ transmission electron microscopy. *Adv. Mater.* **2012**, *24*, 6034–6041. [CrossRef] [PubMed]
51. Zhang, S.; Xu, K.; Jow, T. EIS study on the formation of solid electrolyte interface in Li-ion battery. *Electrochim. Acta* **2006**, *51*, 1636–1640. [CrossRef]
52. Zhang, S.; Xu, K.; Jow, T. Electrochemical impedance study on the low temperature of Li-ion batteries. *Electrochim. Acta* **2004**, *49*, 1057–1061. [CrossRef]

Article

Linear-Polyethyleneimine-Templated Synthesis of N-Doped Carbon Nanonet Flakes for High-performance Supercapacitor Electrodes

Dengchao Xia [1], Junpeng Quan [1], Guodong Wu [1], Xinling Liu [2], Zongtao Zhang [1,*], Haipeng Ji [1], Deliang Chen [1], Liying Zhang [1], Yu Wang [1], Shasha Yi [1], Ying Zhou [1], Yanfeng Gao [1,3,*] and Ren-hua Jin [4,*]

1 School of Material Science and Engineering, Zhengzhou University, No.100 Kexue Ave, Zhengzhou 450001, China
2 College of Chemistry and Materials Science, Shanghai Normal University, No.100 Guilin Rd, Shanghai 200234, China
3 School of Material Science and Engineering, Shanghai University, No.99 Shangda Rd, Shanghai 200444, China
4 Department of Material and Life Chemistry, Kanagawa University, 3-2-7 Rokkakubashi, Kanagawa-ku 221-8686, Japan
* Correspondence: ztzhang@zzu.edu.cn (Z.Z.); yfgao@shu.edu.cn (Y.G.); rhjin@kanagawa-u.ac.jp (R.-h.J.)

Received: 28 July 2019; Accepted: 26 August 2019; Published: 29 August 2019

Abstract: Novel N-doped carbon nanonet flakes (*N*CNFs), consisting of three-dimensional interconnected carbon nanotube and penetrable mesopore channels were synthesized in the assistance of a hybrid catalytic template of silica-coated-linear polyethyleneimine (PEI). Resorcinol-formaldehyde resin and melamine were used as precursors for carbon and nitrogen, respectively, which were spontaneously formed on the silica-coated-PEI template and then annealed at 700 °C in a N_2 atmosphere to be transformed into the hierarchical 3D N-doped carbon nanonetworks. The obtained *N*CNFs possess high surface area (946 m^2 g^{-1}), uniform pore size (2–5 nm), and excellent electron and ion conductivity, which were quite beneficial for electrochemical double-layered supercapacitors (EDLSs). The supercapacitor synthesized from *N*CNFs electrodes exhibited both extremely high capacitance (up to 613 F g^{-1} at 1 A g^{-1}) and excellent long-term capacitance retention performance (96% capacitive retention after 20,000 cycles), which established the current processing among the most competitive strategies for the synthesis of high performance supercapacitors.

Keywords: linear PEI; N-doped carbon nanonet flask (*N*CNFs); template-assisted synthesis; electrochemical properties; supercapacitor

1. Introduction

The critical issues of climate change and rapidly increasing global energy consumption have triggered tremendous research efforts for clean and renewable energy sources, as well as advanced techniques for energy storage and conversion [1–3]. Among the most competitive technologies, such as batteries, fuel cells, and supercapacitors, the electrochemical double-layer supercapacitors (EDLSs) are capturing growing attention, for their wide potential for electric vehicles, digital devices, and pulsing techniques [4,5]. The EDLS is primarily a physical electrostatic behavior in nature, which is generally based on the reversible adsorption of electrolyte ions at the electrode/electrolyte interface [6]. Since the charges are stored on a high surface area without any faradaic reaction involved, long cycle life can be achieved, in addition to the high capacitance, which establishes them as one of the most important emerging energy strategies for consumer electronics, and bridging devices with high energy batteries in hybrid power applications [7].

For EDLSs, carbon based materials are the most commonly used electrode candidates, due to their stable physicochemical properties, fast charging/discharging kinetics, bipolar operational flexibility and low cost [8]. Various kinds of carbon materials, such as activated carbon (AC), mesoporous carbon, carbon nanotubes, and graphene, have been reported as the electrode materials for EDLSs [9–12]. Among these carbon materials, AC have been conventionally used for industrial EDLSs, because of their ultra-high surface area and commercially available mass production. However, the structural shortcomings, for example, unsuitable pore size distribution and limited surface functionality, make AC suffer from low energy density, low conductivity, and slow inner-pore ion diffusion. Appropriately, carbon materials with multiple pore size distribution (micro-, meso-, and macro-pores) are quite necessary to obtain proper energy storage characteristics [13]. This is because although micropores (<2 nm) can provide a large surface area, the infiltration of electrolyte ions into these pores is rather difficult, and the kinetics is quite slow; the transportation of ionic species in macro- (>50 nm) and meso- (2–50 nm) pores, on the other hand, is much easier but their surface to volume ratios are not high enough [14,15].

Three dimensional (3D) hierarchical carbon architectures, such as 3D graphene membranes, 3D carbon nanotubes, and 3D carbon fiber networks, are strongly recommended as promising supercapacitor candidates, due to their special pores, excellent electron conductivity and large, abundant ion pathways [16–19]. For example, densely-packed carbon nanotube (CNT) spherical assemblies demonstrated a specific capacitance of 215 F g^{-1}, which is twice more than that of frequently reported commercial AC electrode. By using polypyrrole (PPy) microsheets as precursors and KOH as the activating agent, 3D hierarchical porous nanostructures with large specific areas of 2870 m^2/g were reported, which demonstrated a high capacity of 318.2 g^{-1} at a current rate of 0.5 A g^{-1}, and excellent retention ability of 95.8% after long-term cycling of more than 10,000 [20]. More recently, by using garlic skin as a source, novel hierarchical porous carbon materials with 3D penetrating pores were successfully synthesized by Han et al., and an ultra-high capacitance of around 380 F g^{-1} at a current density of 3 A g^{-1} accompanied by good rate performance were progressively reported [21]. All these indicated the promising advantages on 3D hierarchical carbon nanostructures for supercapacitor (SC) applications. Moreover, besides these structural related benefits, the electrochemical performance can further be reinforced by surface chemistry modification; e.g., introducing heteroatom doping with N, B, P, or S [22–25], or forming composites with metal oxide nanomaterials, which can provide more competitive advantages for novel SC devices [26–28].

Up to now, various methods, including chemical vapor or physical based deposition, biomass carbonization, hard-templating, etc., have been developed to synthesize 3D hierarchical carbon for supercapacitor applications [29]. On the one hand, most of those methods make it difficult to obtain the proper structures that are required for high performance SC devices; e.g., large surface area, and an efficient pathway for ion and electron transportation; and on the other hand, even for the most frequently reported methods of hard-templating and biomass carbonization, the structural parameters for carbon are usually limited by their raw templates or initial biomass precursors, which can not be feasibly modulated to achieve a better electrochemical performance. Even more, most of these methods are difficult or costly to be applied for large scale production. It is still quite challenging to develop efficient solutions for synthesizing novel hierarchical carbon nanostructures that are suitable for advanced supercapacitors.

Linear polyethyleneimine (L-PEI) refers to an interesting family of molecules, which can adopt a variety of one-dimensional (1D) nanostructures to form hierarchical nano- and micro-structures [30–33]. Distinct structural configurations, like nanoplate, nanowire, mesoporous microsphere, interconnected nanotube, etc., can be easily obtained from this one single polymer, by modulation of the crystalline condition via low temperature solution method. And because there are large amounts of amine groups existing on PEI molecular chains, linear PEI polymers can be used as effective catalysts to prompt the growth of SiO_2 coating layers, enabling them a preponderance for constructing highly ordered 3D nanomaterials [34–36]. Herein, to unravel the flexibility and benefits for PEI as 3D carbon nanostructure constructing template; we reported an efficient method for the preparation of novel 3D hierarchical nitrogen doped carbon nanonet flakes (*N*CNFs), which showed significantly improved electrochemical performance. The *N*CNFs possess large surface area with permeable and interconnected hierarchical pores which facilitate the transmission of electrolyte ion, and nitrogen doped groups in carbon framework contribute effectively on contact with electrolyte solution. Electrochemical measurements indicated a high specific capacity of 613 F g^{-1} at a current density of 1 A g^{-1} (or 259 F g^{-1} at 10 A g^{-1}) and cycling stability after 20,000 cycles at 10 A g^{-1}, which are quite encouraging for applications as high performance SC electrodes.

2. Experimental Section

2.1. Chemicals and Materials

Poly(2-ethyl-2-oxazoline) was bought from Alfa Aesar chemicals, Shanghai. Resorcinol and tetramethoxysilane (TMOS) was purchased from Maclin Biochemical Co., Ltd., Shanghai, China. Melamine, sodium hydroxide, anhydrous ethanol and hydrofluoric acid solution (≥40%) were purchased from Shanghai, China Chemical Regent Co., Ltd. Hydrochloric acid solution (36%–38%) was purchased from Luoyang, China Haohua Chemical Regent Co., Ltd. Formaldehyde solution (≥37%) was purchased from Tianjin, China Jiachen Chemical Factory.

2.2. Synthesis of Linear Polyethyleimine

The synthesis was performed according to the previous report (in Scheme 1). Briefly, 20 g Poly(2-ethyl-2-oxazoline) was dissolved in a 200 mL HCl solution (5 M) and the solution was heated for 12 h under stirring in an oil bath at ca. 100 °C. After cooling to room temperature, a white suspension was obtained. The precipitate was further collected by suction, washed by methanol three times and dried under vacuum. The as-collected product was protonated PEI (PEI-H^+, shown in Scheme 1). 2 g of PEI-H^+ powders were dissolved in 24 mL water and then neutralized by the addition of 5 mL NaOH solution (5 M), which led to the formation of crystalline PEI aggregates. After centrifugation, wet PEI powders were separated and further washed by H_2O three times.

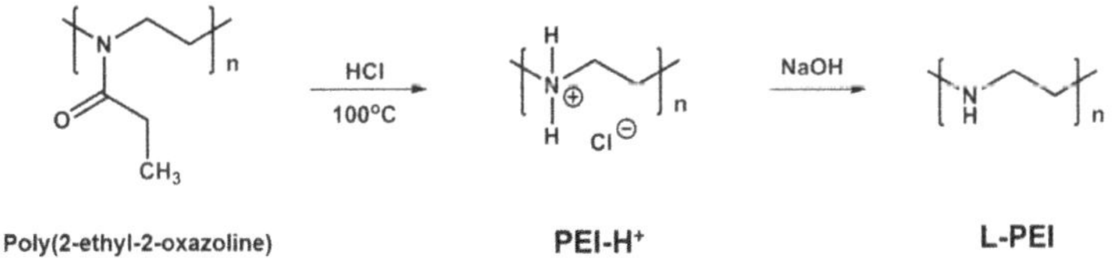

Scheme 1. The synthesis route of linear polyethyleneimine (L-PEI).

2.3. Synthesis of PEI@SiO_2 Nanotubes

Wet PEI powders obtained above were dispersed in 480 mL H_2O and then mixed with 4 mL TMOS. After stirring for 3 h, the suspension was subjected to centrifugation, and the as-collected white precipitates were further washed by H_2O and ethanol and finally dried at 60 °C for 12 h, which produced the powders of PEI@SiO_2 nanotubes.

2.4. Preparation of CNFs and NCNFs

Phenolic resins, which were formed by the polymerization between resorcinol and formaldehyde, were employed as the carbon precursor. Melamine was used as the nitrogen source to synthesize N-doped carbon nanotube networks. Firstly, 0.5 g PEI@SiO_2 powders were dispersed in 50 mL H_2O, and proper amount of resorcinol and formaldehyde were added sequentially to form a suspension, which was subjected to heating with stirring for 24 h in an oil bath at 60 °C. After cooling to room temperature, solid powders were collected by centrifugation, washed with H_2O and ethanol, and dried under vacuum. To synthesize the N-doped samples, melamine in water solution was added just after the reaction of resorcinol and formaldehyde.

The as-obtained solid powders above were transferred into a tube furnace and heated at 700 °C for 1.5 h in flowing nitrogen gas, and both carbon and N-doped carbon coated SiO_2 (SiO_2@C) was formed. The SiO_2 components in SiO_2@C was further removed by HF solution. The samples obtained by adding 0 g, 0.03 g and 0.1 g melamine were denoted as CNFs, *N*CNFs-1 and *N*CNFs-2, respectively.

2.5. Characterization

The transmission electrical microscopy (TEM) measurement was performed by FEI Tecnai G220 (Hillsboro, OR, USA) with an accelerating voltage of 200 kV. SEM characterization was conducted on a JEOL JSM-7500F field-emission scanning electrical microscope (Tokyo, Japan). The samples were coated with a layer of 10-nm-thick platinum film deposited by a JEOL JFC-1600 auto fine coater (Tokyo, Japan) before characterization. Wide-angle X-ray diffraction (XRD: RIGAKU, Karlsruhe, Germany) was conducted at a scanning rate of 5° min^{-1} with CuKα radiation (40 kV, 30 mA). Raman spectra were obtained by using a Confotec MR520 instrument (Graben, Germany)with an excitation laser wavelength of 532 nm, and Si wafers were applied as substrates. X-ray photoelectron spectroscopy (XPS) was conducted with a PHI Quantera SXM (ULVAC-PHI, Kanagawa, Japan) instrument with an AlKα X-ray source, and Ar ion etching was performed before measurement. All of the binding energies were calibrated by referencing to the C1's binding energy (285 eV). The specific surface area and the pore structure were measured by nitrogen sorption by using a JW-BK 112 physisorption analyzer (Beijing, China). The samples were degassed at 120 °C for 2 h before measurement. The specific surface area of samples was calculated by the Brunauer–Emmett–Teller and (BET) method. The pore size distributions were derived from the adsorption of isotherms using the Barrett–Joyner–Halenda (BJH) model.

2.6. Electrical Measurements.

The electrochemical measurements for all samples were characterized with a CHI 660E electrochemical workstation (Shanghai Chenhua, China) in a conventional three-electrode system, and a 6 M KOH aqueous solution was used as the electrolyte. The working electrode was prepared as follows: The active material (80 wt.%), acetylene black (10 wt.%), and polytetrafluoroethylene (PTFE) binder (10 wt.%) were mixed sufficiently with the help of ultrasonic machine. The mixture was pressed into a sheet with a piece of porous nickel net (diameter of 1 cm). The typical loading mass for each electrode is about 0.8–1.5 mg. Platinum and Ag/AgCl (3 M KCl) electrodes were used as the counter and the reference electrodes, respectively. The electrochemical properties of the working electrodes were measured by cyclic voltammetry (CV), galvanostatic charge-discharge (GCD), and electrochemical impedance spectroscopy (EIS). The voltage ranges for the CV and GCD tests were varied from −1 to 0 V. The current density for the galvanostatic measurement were varied from 1 to 20 A g^{-1}. Electrochemical impedance spectroscopy (EIS) was measured in frequent ranges from 10^{-2} Hz to 10^5 Hz at open circuit voltage with a current amplitude of 5 mV. Moreover, to characterize the cycling performance of the samples, galvanostatic measurement at a current density of 10 A g^{-1} was also carried out for 20,000 cycles. The capacitance of the electrode was calculated on the basis of the GCD curve according to the following equation:

$$C_m = \frac{I \times \Delta t}{m \times \Delta V}$$

where C_m (F g^{-1}) is the specific capacitance, I (A) is the discharge current, Δt (s) is the discharge time, m (g) is the mass of active material in the working electrode, and ΔV (V) is the potential window.

3. Results and Discussion

3.1. Structure Design and Characterization

The schematic illustration for the synthesis of *N*CNFs is shown in Figure 1. There are several appealing aspects for the use of linear PEI molecules for constructing the 3D structures: Firstly, as reported before, the L-PEI molecules can crystallize and self-assemble into diverse nano-structures under different crystallization conditions. Secondly, the amine groups on the main chains of PEI can catalyze the quick hydrolysis and condensation of silica source (e.g., TMOS), thus prompt deposition of SiO_2 layers around PEI assembly; thirdly, as the protonated PEI (PEI-H^+) is soluble in H_2O, while the PEI was insoluble at room temperature, the crystallization of PEI can easily be fulfilled by neutralization of PEI-H^+ into PEI crystalline aggregates driven by the addition of alkali solution, which finally transformed into the cross-linked PEI@SiO_2 nanotube networks once mixed with silica sources. Even more, the as-obtained PEI@SiO_2 structures are also effective at prompting the polymerization of resorcinol and formaldehyde to form phenolic resins on the surface of PEI@SiO_2 (Figure 1c), which can be easily turned into SiO_2@C structures after carbonization. To further improve the porosity of 3D *N*CNFs, the inside SiO_2 cores can be removed by HF to form cross-linked hollow nanotube structures (Figure 1d). The above solution-based processing also showed significant changes in powder colors with different compositions, which can be seen in the photograph of Figure 2.

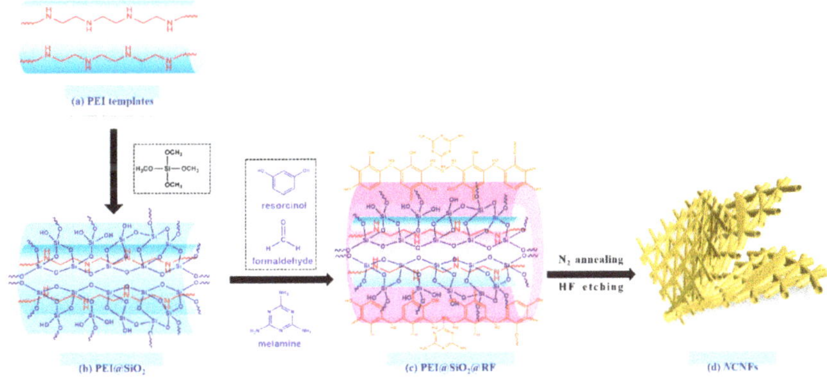

Figure 1. Schematic illustration for the synthesis of N-doped 3-dimensional carbon nanotube networks. (**a**) Linear polyethyleneimine (PEI) template, (**b**) deposition of SiO_2 layer on PEI surface to form the PEI@SiO_2 sample, (**c**) coating of RF onto the PEI@SiO_2 structure to form the PEI@SiO_2@RF sample, and (**d**) the formation of N-doped carbon nanonet flakes (*N*CNFs) and the demonstration of electron and ion transport pathway in carbon nanonet flakes (CNFs).

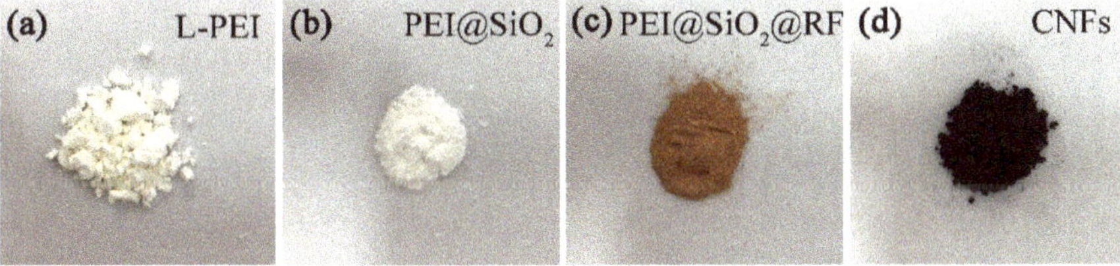

Figure 2. Photographs showing the evolution from linear PEI precursor to the CNFs during the synthesizing process. (**a**) Linear PEI, (**b**) PEI@SiO_2, (**c**) PEI@SiO_2@RF and (**d**) CNFs.

SEM and TEM characterizations were performed to investigate the morphology and structure of the as obtained materials, which are shown in Figure 3. We can see that SiO_2 coated PEI powders

($PEI@SiO_2$) showed flakes-like structure with cross-linked nanofiber networks. After deposition and carbonization of the carbon precursors and further HF etching, the morphologies were mostly retained, remaining in uniform contacts and does not collapse in microscale. The diameter of the nanofiber for $PEI@SiO_2$ was about 20 nm with interconnected mesoscale pores of around 20–30 nm, which were expected to be beneficial for fast electrolyte ion transportation. After further integration with N-doped carbon precursors (melamine), most of the porous structures were still reserved, although the diameter of nanofibers increased by several nanometers. Figure 3e,f showed the TEM and high resolution TEM images for N-doped sample (*N*CNFs-1) after HF etching. The mutually cross-linked nanotubes in the assembled flask structures can be clearly seen. The diameter of the nanotubes for the *N*CNFs-1 is about 15–20 nm with wall thickness of around 3–5 nm, and the length of each nanotube is about dozens to a few hundred nanometers. EDS mapping further confirmed that nitrogen was homogenously distributed around the carbon nanotube networks, which were expected to improve the electron conductivity and surface wettability with electrolytes. The large amounts of pores existing both in the inside of each nanotube and among the large interspaces of adjacent one, may provide enough ion pathways and efficient surface double layer capacity that are required for high performance SC electrodes. Moreover, the special flake-like assembles, which were composed of long nanotubes that interconnected with each other and almost completely spread throughout every flake along the axial direction, could further prompt the formation of stable electron conductive networks. These special and relatively ordered structures are expected to be quite beneficial for high performance SC devices [37], which will be discussed in the following sections.

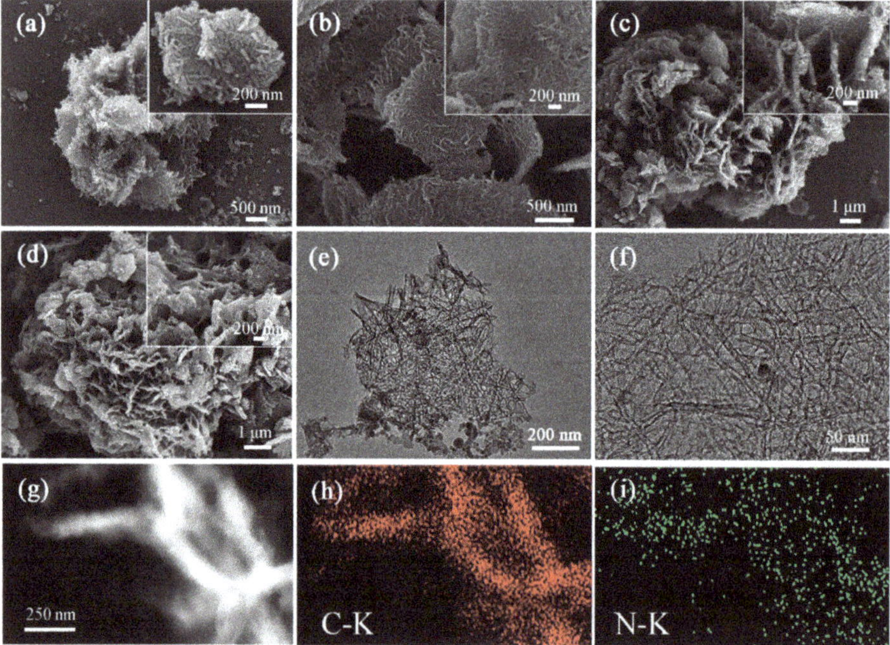

Figure 3. SEM images for $PEI@SiO_2$ (**a**), $SiO_2@C$, (**b**) an N-doped CNFs sample of *N*CNFs-1, (**c**) and *N*CNFs-2 (**d**). (**e**,**f**) Shows the TEM images for sample of *N*CNFs-1 with different magnifications. (**g**–**i**) Shows the TEM dark field image and the corresponding energy dispersive spectrometer (EDS) element mapping results for *N*CNFs-1.

XRD and Raman spectroscopy were further performed to investigate the structure and compositions of CNFs and N-doped samples, which are shown in Figure 4. According to the

XRD results, two broad peaks centered at around 23.4° and 43.0° were observed, which can be assigned to the (002) and (100) diffractions for carbon. Raman spectra further showed the existence of the two typical modes of D (relating with sp^3 hybridization and in plane defects for carbon) and G (relating with sp^2-bonded ordered graphitic carbon), located at around 1363 and 1585 cm^{-1}, respectively, further indicating the formation of carbon in all samples. N_2 adsorption/desorption isotherm measurements were performed to investigate the pore structure for CNFs and nitrogen doped samples, and the results are shown in Figure 4c. All of the N_2 adsorption/desorption isotherm curves exhibited the type-IV profiles, with two typical steep uptakes ($P/P_0 < 0.01$ and $P/P_0 > 0.98$) and hysteresis loops (CNFs and *N*CNFs-1: $0.45 < P/P_0 < 0.97$; *N*CNFs-2: $0.82 < P/P_0 < 0.98$), demonstrating the coexistence of micropores (<2 nm) and mesopores (2–50 nm) in the samples. The pore size distributions of the samples were calculated by using the Barrett–Joyner–Halenda (BJH) method, which is shown in Figure 4d. As can be seen from Figure 4d, all samples demonstrated two distinct peaks in the ranges of 2 to 10 nm (mesopore) and 10 to 200 nm (meso- and macro-pore), respectively, which can be assigned to the inner side pores for each individual carbon nanotube and the void spaces among the interconnected adjacent CNFs, according to the TEM characterizations (Figure 4e,f). The pore contents and their distributions for *N*CNFs-1 were similar with that of CNFs in the range of 2 to 10 nm, although were reduced to some extent especially at the macropore range (>50 nm), which can be explained by the incorporation of additional thin N doped carbon layer on the CNFs. Such peculiar multi-level size distribution is expected to be beneficial for achieving high specific surface area and fast electrolyte ion transferring [38]. Moreover, the calculated specific surface areas were 860, 946, and 365 $m^2\ g^{-1}$, respectively, for CNFs, *N*CNFs-1, and *N*CNFs-2, which was quite attractive among reported 3D hierarchical carbon. The decrease of specific surface area for *N*CNFs-2 can be speculated to owe from the excessive coating of melamine, which suppress both of the mesopores and the macropores, as indicated by the significant decrease of pore volumes for *N*CNFs-2 in Figure 4d.

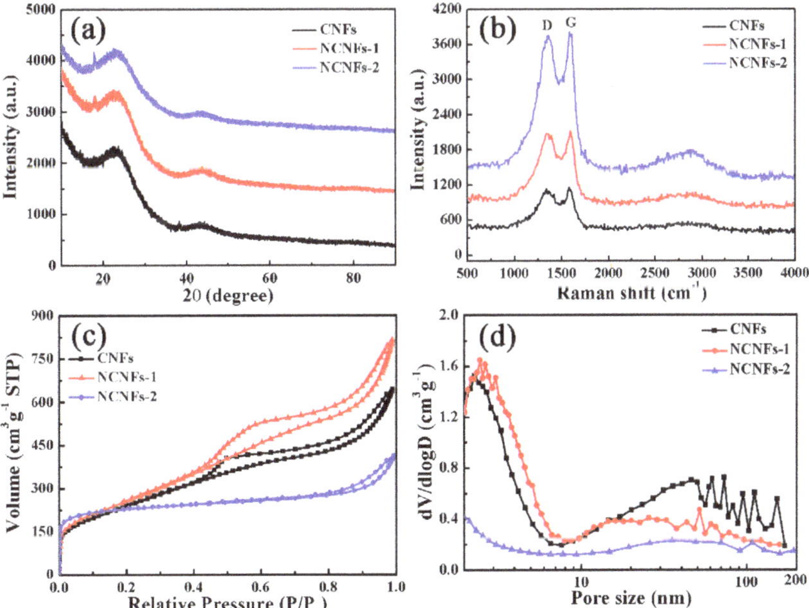

Figure 4. XRD (**a**), Raman (**b**) and nitrogen adsorption/desorption isotherms; (**c**) measurements for CNFs and different N-doped CNF samples. (**d**) Shows the calculated pore size distributions for corresponding samples.

To further investigate the composition of the hierarchical N-doped carbon nanonet flakes, XPS characterizations were performed which are shown in Figure 5. Wide range survey XPS spectra suggested the existence of nitrogen with a content of 1.88 wt.% in the powder after coating of melamine. The forms of nitrogen in carbon frameworks were studied by fitting the high-resolution XPS spectrum for N 1s. There were three significant peaks locating at 398.1 eV, 399.1 eV and 400.5 eV, respectively, for *N*CNFs-1 samples, which can be indexed to the pyridinic-N, pyrrolic-N and the graphitic-N (the configurations of doped N atoms in graphene layer is shown in the insert of Figure 5b). The formation of N doped structures can increase the electron density in carbon to achieve a high electron conductivity, and the defects formed by replacement of C with N can also introduce more active sites to improve the electrochemical performance, both of which are quite beneficial for SC devices.

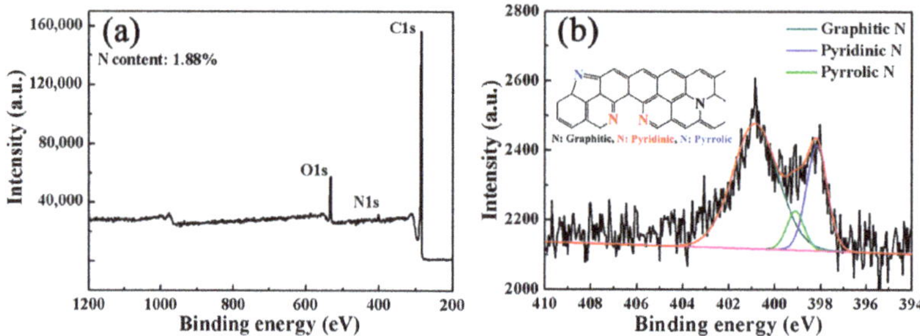

Figure 5. (**a**) X-ray photoelectron spectroscopy (XPS) wide range survey spectrum and (**b**) high-resolution XPS spectrum of N 1s for a sample of *N*CNFs-1 (the inset shows the various configurations of N atoms doped in graphene layer).

3.2. Electrochemical Performance

The electrochemical properties of the CNFs and N-doped CNFs were examined by a three-electrode configuration method, and a 6 M KOH solution was used as the electrolyte. The CV and galvanostatic charge-discharge curves were given in Figure 6. As shown in the CV curves (Figure 6a,c,d), under a fixed potential window of −1 to 0 V (versus Ag/AgCl), all samples indicated the typical response for electric double-layer electrodes, with quasi-rectangular shapes [39]. When a high voltage scan rate, e.g., 500 mV s^{-1}, was applied, the rectangular-like curve was still able to be sustained, demonstrating a relatively fast electron and ion transportation during the charge and discharge process for these samples. Galvanostatic charge–discharge method was also applied to measure the charge and discharge capacitance for SC. The recorded specific capacitances were 461, 613 and 347.5 F g^{-1} (or 322, 351 and 205 F g^{-1}) at a current density of 1 A g^{-1} (2 A g^{-1}), respectively, for CNFs, *N*CNFs-1 and *N*CNFs-2. These values, especially for sample of *N*CNFs-1, are quite high among reports for carbon based EDLSs, which are usually lower than 300 F g^{-1}. Recently, Wu et al. [40]. reported a similar capacitance of around 500 F g^{-1} based on hierarchical 3D carbon electrodes, which consisted of N-doped graphene quantum dots on carbonized MOF and carbon nanotubes hybrid structures. However, less controllability over the structural characteristics and the higher complexity of the synthesizing process can be expected from their reports. Moreover, as PEI refers to an interesting family of molecules, which can adopt a verity of 1D nanostructures to form different hierarchical nano- and micro-structures, and the benefits of feasible interaction chemistry for PEI monocular with SiO_2 and the carbon precursors, the method reported here can provide an effective protocol for building a variety of interesting structures that are quite suitable for supercapacitors.

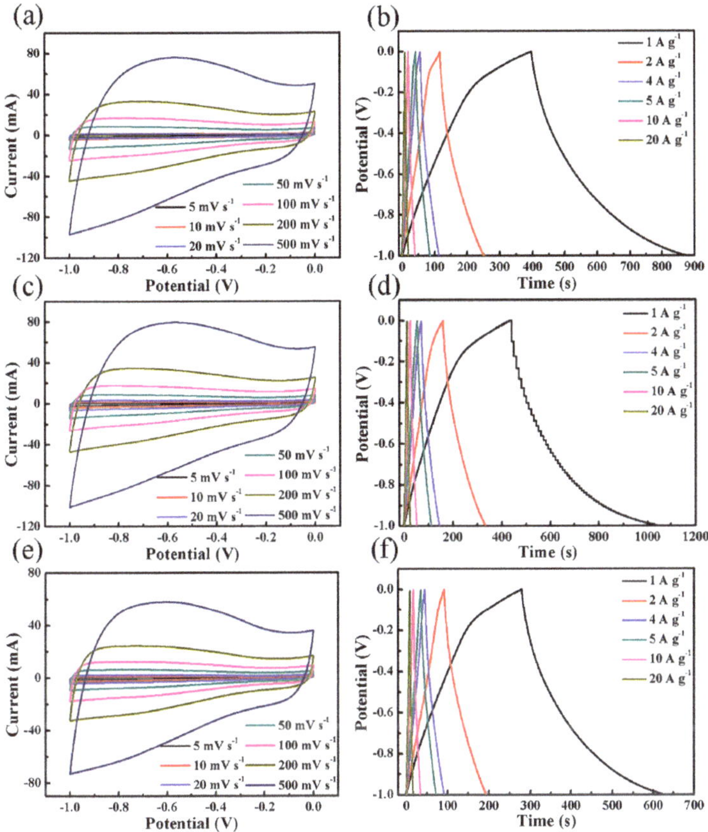

Figure 6. Cyclic voltammetry (CV) curves (**a**,**c**,**e**) at different scan rates and galvanostatic charge discharge curves (**b**,**d**,**f**) at different current densities for CNFs, NCNFs-1 and NCNFs-2, respectively.

To further examine the electrochemical performance, the specific capacitance plots and the long-time cycling measurements for all samples were given, which were shown in Figure 7. We can see that, all samples reached relatively stable capacitance at a high current density range of 5 to 20 A g^{-1}, and more competitively, the NCNF s-1 sample showed a high value of 242 F g^{-1} at 20 A g^{-1}, which was quite high among reports for EDLSs [41–43]. The relatively large drops of the specific capacitance especially at the low current density range (from 1 A g^{-1} to 4 A g^{-1}) could be understood by the suppressed contributions for small pores to the specific capacitance at increased current density as reported by Teng [44] and Daraghmeh [45]. Surface functional groups (see Figure S1 in the Supporting Information) and the related pseudocapacitance [46], as indicated by the slight tailing in the discharge curves in Figure 6, can also cause attenuations of the specific capacitance. However, because of the large amounts of exposed macropores in the hierarchical 3D structures and the excellent electron conductivity derived from the interconnected carbon nanotubes, a relatively high specific capacitance at high current density can still be obtained. Moreover, all samples obtained from PEI templates also showed good cycling stability. As shown in Figure 7b, after 20,000 cycles at a current density of 10 A g^{-1}, specific capacitances of 213 F g^{-1} and 232 F g^{-1} can be obtained for CNFs and NCNFs-1, respectively, with high retention rate of both 95%. Additionally, to investigate the practical benefits of the as-synthesized NCNFs-1 sample, a symmetric two-electrode soft pack device was also constructed (see Figure S2 in the supporting information). The device demonstrated excellent electrochemical performance with high specific capacitances of up to 313.6 F g^{-1} at 0.5 A g^{-1} and 262.4 F g^{-1} at 2 A g^{-1}, which were quite attractive among the reported two-electrode supercapacitor devices. Schematic illustration of the local electron and ion pathway for the 3D hierarchical NCNFs is shown in inset of Figure 7b. The excellent electrochemical performance can be attributed to the following reasons: Firstly, the vast specific surface area as measured by BET can allow large amounts of charge accumulations on the surface or interface between electrodes and electrolyte; secondly, there were a great many stable interconnected and penetrable mesoscale pores (as shown in the SEM and TEM images), which can effectively facilitate the electron and ion transportation for fast electrode dynamics; moreover, the incorporation of nitrogen dopant in carbon nanotubes can significantly enhance the electric conductivity and electrolyte solution wettability.

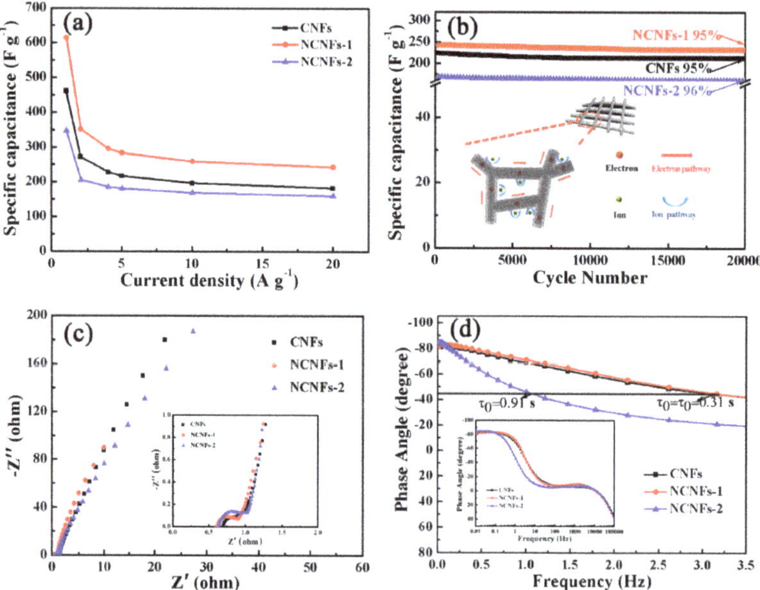

Figure 7. (**a**) The specific capacitance calculated by galvanostatic charge-discharge curves at different current densities ranging from 1 A g^{-1} to 20 A g^{-1}, (**b**) cycling stability at 10 A g^{-1} for NCNFs-1, CNFs and NCNFs-2 (the inset in (**b**) shows the schematic illustration for electron and ion pathway of the 3D hierarchical nitrogen doped carbon nanonet flakes). (**c**) Nyquist plots. (Z′: real impedance, Z″: imaginary impedance. And the inset shows a partial enlarged view in high frequency range). (**d**) Bode plots of phase angle relate to frequency in the low frequency range (the inset shows the normal bode plot).

To further explorer the influence of hierarchical pores and the interconnected flake-like structures on the properties of the working electrodes, electrochemical impedance spectroscopy (EIS) was conducted, and the results are shown in Figure 7c,d. As can be seen from Figure 7c, the Nyquist plots, for CNFs, NCNFs-1 and NCNFs-2, constituted a semicircle in the high frequency region and a straight line in the low frequency region. The nearly vertical profile in the low frequency region for all samples indicated the desired electrical double-layer behavior. As reported before, the semicircle at a high frequency region is related with the electronic resistance between the electrode materials and the electrolyte [47,48]. The resistance for NCNFs-1 is 0.73 Ω, which is slightly lower than that for the CNFs and NCNFs-2 (details can be seen in the enlarged view in Figure 7c), indicating a higher electron conductivity for NCNFs-1. Bode plots of phase angle versus the applied frequency were further performed, which are shown in Figure 7d. We can see that the phase angles of all samples were located around 82° to 85° at the low frequency of zero, agreeing with ideal capacitive behavior for EDLS devices [49]. Moreover, the calculated characteristic frequencies f_0, defined as the frequency at phase angle of −45°, were 3.2 Hz, 3.2 Hz and 1.1 Hz for *N*CNFs-1, CNFs and *N*CNFs-2, respectively, corresponding to a time constant τ_0, defined as $\tau_0 = {f_0}^{-1}$, of 0.31 s, 0.31 s and 0.91 s. The high phase angle and short time constant indicated a faster frequency response and enhanced ionic transport rate for *N*CNFs-1 than that of CNFs and *N*CNFs-2 samples, which were quite beneficial for ion transport dynamic and high performance EDLS devices.

4. Conclusions

We have demonstrated a facile way for the synthesis of three-dimensional N-doped carbon nanotube flake structures by using linear PEI as catalytic template. The as-prepared N carbon materials exhibited high specific surface area, a stable interconnect nanotube network and 3D penetrable micro to macro scale hierarchical pores, which were beneficial for achieving high performance ELDSs. For nitrogen doped sample of *N*CNFs-1, ultra-high specific capacitances of 613 F g^{-1} at 1 A g^{-1} and 242 F g^{-1} at 10 A g^{-1}, and excellent cycle ability with 95% retention over 20,000 cycles at 10 A g^{-1}, was obtained, indicating the current processing is of great benefit for developing high performance electrodes for supercapacitors.

Supplementary Materials: The following are available online at http://www.mdpi.com/2079-4991/9/9/1225/s1, Figure S1: The fitted high-resolution XPS spectrum of C 1s for the sample of *N*CNFs-1, Figure S2: The electrochemical performance of *N*CNFs-1 in a two-electrode system.

Author Contributions: D.X. and J.Q. contributed equally to this paper. Z.Z., Y.G. and X.L. designed the experiment. D.X., J.Q. and G.W. performed the experiments. L.Z. and S.Y. performed the electrochemical measurements. Z.Z., Y.G. and R.-h.J. designed the framework of the manuscript. D.X., J.Q., Z.Z., X.L., R.-h.J. and H.J. wrote the paper, D.C., Y.W. and Y.Z. provided some guidance for the experiment, and all authors participated in the discussion of the paper.

Funding: This study was supported in part by funding from the National Natural Science Foundation of China (NSFC, contract numbers: 51502268, 51325203 and 51574205), the Excellent Young Teacher Development Foundation of Zhengzhou University (contract cumber: 1421320050), and the Key Science and Technology Research Projects of Henan Provincial Education Department (contract number: 14B430023).

Conflicts of Interest: The authors declare no conflict of interest.

References

1. Chen, T.; Dai, L. Carbon nanomaterials for high-performance supercapacitors. *Mater. Today* **2013**, *16*, 272–280. [CrossRef]
2. Zhou, J.; Gao, Y.; Zhang, Z.; Luo, H.; Cao, C.; Chen, Z.; Dai, L.; Liu, X. VO2 thermochromic smart window for energy savings and generation. *Sci. Rep.* **2013**, *3*, 3029. [CrossRef] [PubMed]
3. Cui, Y.; Ke, Y.; Liu, C.; Chen, Z.; Wang, N.; Zhang, L.; Zhou, Y.; Wang, S.; Gao, Y.; Long, Y. Thermochromic VO2 for Energy-Efficient Smart Windows. *Joule* **2018**, *2*, 1707–1746. [CrossRef]
4. Wang, Y.; Song, Y.; Xia, Y. Electrochemical capacitors: Mechanism, materials, systems, characterization and applications. *Chem. Soc. Rev.* **2016**, *45*, 5925–5950. [CrossRef] [PubMed]
5. Han, J.; Xu, G.; Ding, B.; Pan, J.; Dou, H.; Macfarlane, D.R. Porous nitrogen-doped hollow carbon spheres derived from polyaniline for high performance supercapacitors. *J. Mater. Chem. A* **2014**, *2*, 5352–5357. [CrossRef]
6. Liu, C.; Yu, Z.; Neff, D.; Zhamu, A.; Jang, B.Z. Graphene-Based Supercapacitor with an Ultrahigh Energy Density. *Nano Lett.* **2010**, *10*, 4863–4868. [CrossRef] [PubMed]
7. Ji, H.; Zhao, X.; Qiao, Z.; Jung, J.; Zhu, Y.; Lu, Y.; Zhang, L.L.; Macdonald, A.H.; Ruoff, R.S. Capacitance of carbon-based electrical double-layer capacitors. *Nat. Commun.* **2014**, *5*, 3317. [CrossRef] [PubMed]
8. Frackowiak, E.; Abbas, Q.; Béguin, F. Carbon/carbon supercapacitors. *J. Energy Chem.* **2013**, *22*, 226–240. [CrossRef]
9. Ahmed, S.; Ahmed, A.; Rafat, M. Supercapacitor performance of activated carbon derived from rotten carrot in aqueous, organic and ionic liquid based electrolytes. *J. Saudi Chem. Soc.* **2018**, *22*, 993–1002. [CrossRef]

10. Tang, D.; Hu, S.; Dai, F.; Yi, R.; Gordin, M.L.; Chen, S.; Song, J.; Wang, D. Self-Templated Synthesis of Mesoporous Carbon from Carbon Tetrachloride Precursor for Supercapacitor Electrodes. *ACS Appl. Mater. Interfaces* **2016**, *8*, 6779–6783. [CrossRef] [PubMed]
11. Wu, G.; Tan, P.; Wang, D.; Li, Z.; Peng, L.; Hu, Y.; Wang, C.; Zhu, W.; Chen, S.; Chen, W. High-performance Supercapacitors Based on Electrochemical-induced Vertical-aligned Carbon Nanotubes and Polyaniline Nanocomposite Electrodes. *Sci. Rep.* **2017**, *7*, 43676. [CrossRef] [PubMed]
12. Ke, Q.; Wang, J. Graphene-based materials for supercapacitor electrodes—A review. *J. Mater.* **2016**, *2*, 37–54. [CrossRef]
13. Kondrat, S.; Perez, C.; Presser, V.; Gogotsi, Y.; Kornyshev, A.A. Effect of pore size and its dispersity on the energy storage in nanoporous supercapacitors. *Energy Environ. Sci.* **2012**, *5*, 6474. [CrossRef]
14. Vaquero, S.; Diaz, R.; Anderson, M.; Palma, J.; Marcilla, R. Insights into the influence of pore size distribution and surface functionalities in the behaviour of carbon supercapacitors. *Electrochim. Acta* **2012**, *86*, 241–247. [CrossRef]
15. Zhang, L.; Yang, X.; Zhang, F.; Long, G.; Zhang, T.; Leng, K.; Zhang, Y.; Huang, Y.; Ma, Y.; Zhang, M.; et al. Controlling the Effective Surface Area and Pore Size Distribution of sp2Carbon Materials and Their Impact on the Capacitance Performance of These Materials. *J. Am. Chem. Soc.* **2013**, *135*, 5921–5929. [CrossRef] [PubMed]
16. Yu, Z.; Tetard, L.; Zhai, L.; Thomas, J. Supercapacitor electrode materials: Nanostructures from 0 to 3 dimensions. *Energy Environ. Sci.* **2015**, *8*, 702–730. [CrossRef]
17. Liang, C.; Hong, K.; Guiochon, G.A.; Mays, J.W.; Dai, S. Synthesis of a Large-Scale Highly Ordered Porous Carbon Film by Self-Assembly of Block Copolymers. *Angew. Chem.* **2004**, *116*, 5909–5913. [CrossRef]
18. Jiang, H.; Lee, P.S.; Li, C. 3D carbon based nanostructures for advanced supercapacitors. *Energy Environ. Sci.* **2013**, *6*, 41–53. [CrossRef]
19. Zhao, J.; Jiang, Y.; Fan, H.; Liu, M.; Zhuo, O.; Yang, L.; Ma, Y.; Hu, Z.; Wang, X.; Wu, Q. Porous 3D Few-Layer Graphene-like Carbon for Ultrahigh Power Supercapacitors with Well-Defined Structure-Performance Relationship. *Adv. Mater.* **2017**, *29*, 1604569. [CrossRef] [PubMed]
20. Qie, L.; Chen, W.; Xu, H.; Xiong, X.; Jiang, Y.; Zou, F.; Hu, X.; Xin, Y.; Zhang, Z.; Huang, Y. Synthesis of functionalized 3D hierarchical porous carbon for high-performance supercapacitors. *Energy Environ. Sci.* **2013**, *6*, 2497–2504. [CrossRef]
21. Guo, Z.; Feng, X.; Li, X.; Zhang, X.; Peng, X.; Song, H.; Fu, J.; Ding, K.; Huang, X.; Gao, B. Nitrogen Doped Carbon Nanosheets Encapsulated in situ Generated Sulfur Enable High Capacity and Superior Rate Cathode for Li-S Batteries. *Front. Chem.* **2018**, *6*, 429. [CrossRef] [PubMed]
22. Xiao, Q.; Shen, J.; Li, B.; Dai, F.; Yang, L.; Zhang, C.; Cai, M. Nitrogen-doped activated carbon for a high energy hybrid supercapacitor. *Energy Environ. Sci.* **2016**, *9*, 102–106.
23. Zhao, Z.; Xie, Y. Electrochemical supercapacitor performance of boron and nitrogen co-doped porous carbon nanowires. *J. Power Sources* **2018**, *400*, 264–276. [CrossRef]
24. Yi, J.; Qing, Y.; Wu, C.; Zeng, Y.; Wu, Y.; Lu, X.; Tong, Y. Lignocellulose-derived porous phosphorus-doped carbon as advanced electrode for supercapacitors. *J. Power Sources* **2017**, *351*, 130–137. [CrossRef]
25. Chen, H.; Yu, F.; Wang, G.; Chen, L.; Dai, B.; Peng, S. Nitrogen and Sulfur Self-Doped Activated Carbon Directly Derived from Elm Flower for High-Performance Supercapacitors. *ACS Omega* **2018**, *3*, 4724–4732. [CrossRef] [PubMed]
26. Jiang, H.; Ma, J.; Li, C. Mesoporous carbon incorporated metal oxide nanomaterials as supercapacitor electrodes. *Adv. Mater.* **2012**, *24*, 4197–4202. [CrossRef] [PubMed]
27. Wang, Y.; Guo, J.; Wang, T.; Shao, J.; Wang, D.; Yang, Y.-W. Mesoporous Transition Metal Oxides for Supercapacitors. *Nanomaterials* **2015**, *5*, 1667–1689. [CrossRef] [PubMed]
28. Sharma, V.; Singh, I.; Chandra, A. Hollow nanostructures of metal oxides as next generation electrode materials for supercapacitors. *Sci. Rep.* **2018**, *8*, 1307. [CrossRef]
29. Liu, Y.; Chen, J.; Cui, B.; Yin, P.; Zhang, C. Design and Preparation of Biomass-Derived Carbon Materials for Supercapacitors: A Review. *Carbon* **2018**, *4*, 53. [CrossRef]
30. Lungu, C.N.; Diudea, M.V.; Putz, M.V.; Grudziński, I.P. Linear and Branched PEIs (Polyethylenimines) and Their Property Space. *Int. J. Mol. Sci.* **2016**, *17*, 555. [CrossRef]
31. Wang, F.; Sun, L.; Chen, C.; Chen, Z.; Zhang, Z.; Wei, G.; Jiang, X. Polyethyleneimine templated synthesis of hierarchical SAPO-34 zeolites with uniform mesopores. *RSC Adv.* **2014**, *4*, 46093–46096. [CrossRef]

32. Zakeri, A.; Kouhbanani, M.A.J.; Beheshtkhoo, N.; Beigi, V.; Mousavi, S.M.; Hashemi, S.A.R.; Zade, A.K.; Amani, A.M.; Savardashtaki, A.; Mirzaei, E.; et al. Polyethylenimine-based nanocarriers in co-delivery of drug and gene: A developing horizon. *Nano Rev. Exp.* **2018**, *9*, 1488497. [CrossRef] [PubMed]
33. Jin, R.-H.; Yao, D.-D.; Levi, R.T. Biomimetic Synthesis of Shaped and Chiral Silica Entities Templated by Organic Objective Materials. *Chem. A Eur. J.* **2014**, *20*, 7196–7214. [CrossRef] [PubMed]
34. Liu, X.-L.; Moriyama, K.; Gao, Y.-F.; Jin, R.-H. Polycondensation and carbonization of phenolic resin on structured nano/chiral silicas: Reactions, morphologies and properties. *J. Mater. Chem. B* **2016**, *4*, 626–634. [CrossRef]
35. Jäger, M.; Schubert, S.; Ochrimenko, S.; Fischer, D.; Schubert, U.S. Branched and linear poly(ethylene imine)-based conjugates: Synthetic modification, characterization, and application. *Chem. Soc. Rev.* **2012**, *41*, 4755. [CrossRef] [PubMed]
36. Yuan, J.-J.; Jin, R.-H. Temporally and spatially controlled silicification for self-generating polymer@silica hybrid nanotube on substrates with tunable film nanostructure. *J. Mater. Chem.* **2012**, *22*, 5080. [CrossRef]
37. Wu, J.; Shi, X.; Song, W.; Ren, H.; Tan, C.; Tang, S.; Meng, X. Hierarchically porous hexagonal microsheets constructed by well-interwoven MCo_2S_4 (M = Ni, Fe, Zn) nanotube networks via two-step anion-exchange for high-performance asymmetric supercapacitors. *Nano Energy* **2018**, *45*, 439–447. [CrossRef]
38. Fang, M.; Chen, Z.; Tian, Q.; Cao, Y.; Wang, C.; Liu, Y.; Fu, J.; Zhang, J.; Zhu, L.; Yang, C.; et al. Synthesis of uniform discrete cage-like nitrogen-doped hollow porous carbon spheres with tunable direct large mesoporous for ultrahigh supercapacitive performance. *Appl. Surf. Sci.* **2017**, *425*, 69–76. [CrossRef]
39. Liu, C.; Wang, J.; Li, J.; Zeng, M.; Luo, R.; Shen, J.; Sun, X.; Han, W.; Wang, L. Synthesis of N-Doped Hollow-Structured Mesoporous Carbon Nanospheres for High-Performance Supercapacitors. *ACS Appl. Mater. Interfaces* **2016**, *8*, 7194–7204. [CrossRef]
40. Li, Z.; Liu, X.; Wang, L.; Bu, F.; Wei, J.; Pan, D.; Wu, M. Hierarchical 3D All-Carbon Composite Structure Modified with N-Doped Graphene Quantum Dots for High-Performance Flexible Supercapacitors. *Small* **2018**, *14*, e1801498. [CrossRef]
41. Wang, G.; Zhang, L.; Zhang, J. A review of electrode materials for electrochemical supercapacitors. *Chem. Soc. Rev.* **2012**, *41*, 797–828. [CrossRef] [PubMed]
42. Wang, F.; Wu, X.; Yuan, X.; Liu, Z.; Zhang, Y.; Fu, L.; Zhu, Y.; Zhou, Q.; Wu, Y.; Huang, W. Latest advances in supercapacitors: From new electrode materials to novel device designs. *Chem. Soc. Rev.* **2017**, *46*, 6816–6854. [CrossRef] [PubMed]
43. Chen, X.; Paul, R.; Dai, L. Carbon-based supercapacitors for efficient energy storage. *Natl. Sci. Rev.* **2017**, *4*, 453–489. [CrossRef]
44. Teng, S.; Siegel, G.; Wang, W.; Tiwari, A. Carbonized Wood for Supercapacitor Electrodes. *ECS Solid State Lett.* **2014**, *3*, M25–M28. [CrossRef]
45. Daraghmeh, A.; Hussain, S.; Saadeddin, I.; Servera, L.; Xuriguera, E.; Cornet, A.; Cirera, A. A Study of Carbon Nanofibers and Active Carbon as Symmetric Supercapacitor in Aqueous Electrolyte: A Comparative Study. *Nanoscale Res. Lett.* **2017**, *12*, 639. [CrossRef] [PubMed]
46. Cao, H.; Peng, X.; Zhao, M.; Liu, P.; Xu, B.; Guo, J. Oxygen functional groups improve the energy storage performances of graphene electrochemical supercapacitors. *RSC Adv.* **2018**, *8*, 2858–2865. [CrossRef]
47. Yang, Z.; Chabi, S.; Xia, Y.; Zhu, Y. Preparation of 3D graphene-based architectures and their applications in supercapacitors. *Prog. Nat. Sci. Mater. Int.* **2015**, *25*, 554–562. [CrossRef]
48. Zhang, M.; Wang, G.; Lu, L.; Wang, T.; Xu, H.; Yu, C.; Li, H.; Tian, W. Improving the electrochemical performances of active carbon-based supercapacitors through the combination of introducing functional groups and using redox additive electrolyte. *J. Saudi Chem. Soc.* **2018**, *22*, 908–918. [CrossRef]
49. Purkait, T.; Singh, G.; Kumar, D.; Singh, M.; Dey, R.S. High-performance flexible supercapacitors based on electrochemically tailored three-dimensional reduced graphene oxide networks. *Sci. Rep.* **2018**, *8*, 640. [CrossRef]

Article

Mixed-Phase MnO_2/N-Containing Graphene Composites Applied as Electrode Active Materials for Flexible Asymmetric Solid-State Supercapacitors

Hsin-Ya Chiu and Chun-Pei Cho *

Department of Applied Materials and Optoelectronic Engineering, National Chi Nan University, Nantou 54561, Taiwan; s103328505@mail1.ncnu.edu.tw

* Correspondence: cpcho@ncnu.edu.tw

Received: 13 October 2018; Accepted: 5 November 2018; Published: 8 November 2018

Abstract: MnO_2/N-containing graphene composites with various contents of Mn were fabricated and used as active materials for the electrodes of flexible solid-state asymmetric supercapacitors. By scanning electron microscopes (SEM), transmission electron microscope (TEM), energy-dispersive X-ray spectroscopy (EDS), X-ray photoelectron spectrometer (XPS), fourier-transform infrared spectroscopy (FTIR) and Raman spectra, the presence of MnO_2 and N-containing graphene was verified. The MnO_2 nanostructures decorated on the N-containing graphene were of α- and γ-mixed phases. N-containing graphene was found to reduce the charge transfer impedance in the high-frequency region at the electrode/electrolyte interface (R_{CT}) due to its good conductivity. The co-existence of MnO_2 and N-containing graphene led to a more reduced R_{CT} and improved charge transfer. Both the mass loading and content of Mn in an active material electrode were crucial. Excess Mn caused reduced contacts between the electrode and electrolyte ions, leading to increased R_{CT}, and suppressed ionic diffusion. When the optimized mass loading and Mn content were used, the 3-NGM1 electrode exhibiting the smallest R_{CT} and a lower ionic diffusion impedance was obtained. It also showed a high specific capacitance of 638 $F \cdot g^{-1}$ by calculation from the cyclic voltammetry (CV) curves. The corresponding energy and power densities were 372.7 $Wh \cdot kg^{-1}$ and 4731.1 $W \cdot kg^{-1}$, respectively. The superior capacitance property arising from the synergistic effect of mixed-phase MnO_2 and N-containing graphene had permitted the composites promising active materials for flexible solid-state asymmetric supercapacitors. Moreover, the increase of specific capacitance was found to be more significant by the pseudocapacitive MnO_2 than N-containing graphene.

Keywords: MnO_2; N-containing graphene; composite; active material; specific capacitance; asymmetric supercapacitor

---

1. Introduction

The rapid development of portable and wearable consumer electronics results in the demand for high-performance energy-storage devices [1–7]. Solid-state supercapacitors (SSCs) have thus attracted growing attention, due to their reversibility, safe operation without the use of any liquid electrolyte, longer cycling stability than batteries, and higher power and energy densities than conventional capacitors [1,2,8,9]. Further improvement in energy density and searching for economic flexible current collectors are regarded as the two major challenges when pushing the technology of SSCs forward. Lots of effort have been made to investigate electrode active materials, to achieve higher specific capacitance. Carbonaceous materials such as activated carbon, carbon fibers, carbon nanotubes, graphene (G), and graphene oxides, etc., can be used as the active materials for electrodes of SSCs due to low cost, high conductivity, and chemical stability [10–16]. However, they usually show the drawback of lower specific capacitance, as compared to manganese oxide (MnO_2) [17,18]. The theoretical specific capacitance of porous carbon materials is only 250 F·g^{-1} when the specific surface area is 1000 m^2·g^{-1} [19]. Modification of pure carbon, like mixing activated carbon with conductive additives to obtain active materials with a high surface area and better contact for more efficient electrodes, was reported to improve the conductivity and capacitances of supercapacitors (SCs) [17]. Addition of heteroatom(s) or a pseudocapacitive material in G to fabricate composites was also conducive to the performance of SCs. Nitrogen (N) has a high electronegativity and lone pair of electrons, which can be in conjugation with the π electrons of G to enhance conductivity and electron transport. So, compared to pure G, N-containing G (NG) would be more favorable to the specific capacitance of SSCs.

Compared with carbonaceous materials, transition metal oxides and hydroxides (RuO_2, MnO_2, $Ni(OH)_2$, Fe_2O_3, etc.) are deemed as more promising active materials for electrode applications to SSCs with large specific capacitances and high energy densities because the diverse oxidation states of the transition metals permit effective charge transfer [20–24]. Among them, MnO_2 has attracted considerable attention due to its low cost, natural abundance, environmentally benign nature, and high theoretical specific capacitance (1370 F·g^{-1}) [25–27]. Its pseudocapacitive characteristic can be attributed to the single electron transfer in the Mn^{3+}/Mn^{4+} redox system [28]. Nevertheless, its poor conductivity and low ion diffusion constant may suppress the further progress [29,30]. The stability and potential application may be hindered by its problematic dissolution in electrolytes. To address these issues, the combination of MnO_2 nanostructures with conductive carbonaceous materials to form hybrid structures has been adopted [31–39]. Some studies have focused on the development of MnO_2-G composites, which took advantages of the high capacitance of MnO_2 and the high conductivity of G simultaneously.

Despite the advantages of good electrochemical stability, high conductivity, and large specific surface area, G nanosheets tend to restack during the formation of solid materials due to the strong π-π interactions. The aggregation reduces the accessible surface area for adsorption and desorption of electrolyte ions, which would finally result in a small specific capacitance. The high conductivity and unique surface characteristics of the single-layer G nanosheets can be thereby lost [40]. Thus, suppression of the aggregation would greatly optimize the electrochemical properties of electrodes. One feasible method is to anchor transition metal oxides to the surface of G, working as spacers to separate adjacent G nanosheets. Transition metal oxides such as RuO_2, NiO, CoO_x, and MnO_2 are appropriate candidates since they have been considered as extensively explored pseudocapacitor's electrode materials showing high theoretical specific capacitance [41–44]. Introducing porous MnO_2 nanostructures into G nanosheets would suppress the aggregation. The specific surface area of G was then increased, and more electrical conduction pathways were provided. The relatively higher gravimetric capacitances had been demonstrated for a variety of MnO_2-G composites with low mass loadings, whereas some literature highlighted the importance of fabricating composites with higher mass loadings [45–50]. The optimization of mass loading would remain an important challenge.

The asymmetric SCs consisted of a carbonaceous negative electrode and a MnO_2-based positive electrode offered enlarged operation voltage windows and thus improved power energy properties, compared with symmetric SCs [51]. For example, MnO_2 nanostructures grown on activated carbon by a wet chemical reaction process was used as the positive electrode, which exhibited a high specific capacitance (345.1 F·g^{-1} at 0.5 A·g^{-1}) and excellent cycle stability. It was assembled with an activated carbon negative electrode to fabricate asymmetric SCs, which showed high energy density of 31.0 Wh·kg^{-1} at a power density of 500.0 W·kg^{-1} [19]. Layered δ-MnO_2 on N-doped G obtained by a hydrothermal approach was used as the cathode to improve the conductivity and present a high specific capacitance of about 305 F·g^{-1} at a scan rate of 5 mV·s^{-1}. When it was assembled with an activated carbon anode using a gel electrolyte to fabricate flexible asymmetric SSCs (ASSCs), a maximum energy density of 3.5 mWh·cm^{-3} at a power density of 0.019 W·cm^{-3} was achieved [52]. Despite the progress, most MnO_2-based SSCs did not exceed the energy density of lead acid batteries. Lots of efforts have been made to ASSCs with various electrode combinations [53–55]. Consequently, it is crucial to develop new active materials for more efficient electrodes applied to SCs.

In this study, MnO_2/NG composites with various contents of Mn were fabricated by a hydrothermal approach and used as the electrode active materials for flexible ASSCs. Graphite paper on polyimide (PI) was employed as the soft substrate. NG can enhance the conductivity of composites and efficiently reduce the interfacial impedance. It can also serve as a better template for inducing the growth of MnO_2 nanostructures than G. By the synergistic effect of MnO_2 and NG, the specific capacitance, energy, and power densities were significantly improved. The MnO_2 in the composites was found to be mixed phases containing γ-MnO_2 and α-MnO_2. The impacts of mass loading and the content of Mn on the capacitance parameters were also explored. The 3-NGM1 electrode with the most appropriate Mn content and mass loading of active material exhibited a high specific capacitance of 258 $F \cdot g^{-1}$ at a current density of 1 $A \cdot g^{-1}$. By calculating the cyclic voltammetry (CV) results, it had a superior specific capacitance of 638 $F \cdot g^{-1}$. The corresponding energy and power densities were 372.7 $Wh \cdot kg^{-1}$ and 4731.1 $W \cdot kg^{-1}$, respectively. The ongoing work regarding flexible ASSCs will be designed as using G as the negative electrode and a MnO_2/NG composite as the positive electrode by employing a solid gel electrolyte. It is perceived that the optimized conditions of electrodes will lead to more enhanced capacitive behavior and cycle stability of the flexible ASSCs.

2. Experimental

2.1. Preparation of G

Graphite oxide (GO) was synthesized by the modified Hummers' method using graphite powder [56]. A graphite oxidation procedure was executed before the synthesis of GO [57,58]. 4 g of graphite powder were added into a solution composed of 2 g of potassium persulfate ($K_2S_2O_8$), 2 g of phosphorus pentoxide (P_2O_5), and 30 mL of conc. sulfuric acid (H_2SO_4). The mixture solution was heated to 80 °C under continuous stirring for 6 h. When it was cooled down to room temperature, rinse with deionized (DI) water was performed repeatedly by centrifugation until the neutral pH level was achieved. Afterward, 4 g of the pre-oxidized graphite powder were added into 100 mL of conc. H_2SO_4 solution in an ice bath. Then 12 g of potassium permanganate (KM_nO_4) were slowly added at 35 °C. The stirring was continued for 2 h until the color of the mixture turned to dark brown. Subsequently, a solution containing 200 mL of DI water and 40 mL of hydrogen peroxide (H_2O_2, 30 vol% in water) was added slowly while a violent chemical reaction occurred. A yellow-brown intermediate was produced when the reaction was completed, which was then put in a dilute aqueous hydrochloric acid (HCl) solution to remove metal ions. After ultrasonication for 1 h, rinse with DI water was repeatedly performed by centrifugation until the neutral pH level was achieved to obtain the GO powder.

20 mg of GO were added in 100 mL of DI water to prepare the GO solution. After ultrasonication for 2 h to make better dispersion of GO, the suspension was transferred to an autoclave, which was placed in a furnace for the hydrothermal process at 200 °C for 2 h. When cooling down to room temperature, the product was collected by filtration, and then dried at 80 °C for 12 h. After grinding, the powder of G was acquired [59].

2.2. Preparation of NG Composites

55 mg of G were mixed with 8.6 mL of ammonia hydroxide solution (28 vol%~30 vol%) in 70 mL of DI water. After ultrasonication for 2 h to make better dispersion of G, the suspension was transferred to an autoclave, which was placed in a furnace for the hydrothermal process at 140 °C for 6 h. When cooling down to room temperature, the product was repeatedly rinsed by DI water until the neutral pH level was achieved, and then collected by centrifugation. After dried at 80 °C for 12 h and grinding, the NG powder was obtained [52].

2.3. Preparation of NG/MnO_2 (NGM) Composites

55 mg of G were mixed with 8.6 mL of ammonia hydroxide solution (28 vol%~30 vol%) in 70 mL of DI water. After ultrasonication for 2 h to make better dispersion of G, the suspension was transferred

to an autoclave, which was placed in a furnace for the hydrothermal process at 140 °C for 6 h. When cooling down to the room temperature, five different weights of potassium permanganate ($KMnO_4$) were added, respectively, to prepare the mixtures containing 8.9 mM, 17.8 mM, 26.7 mM, 35.6 mM, and 44.5 mM of $KMnO_4$ solutions. After ultrasonication for 30 min, every mixture was transferred back to the autoclave, which was then placed in the furnace for another hydrothermal process at 160 °C for 2 h. When cooling down to room temperature, every mixture was taken out and repeatedly rinsed by DI water until the neutral pH level was achieved, and then collected by centrifugation. After dried at 80 °C for 12 h and grinding, five NGM composites with various contents of Mn were obtained [52]. They were named as x-NGM, in which x was 1, 2, 3, 4, and 5, respectively, to represent the five $KMnO_4$ concentrations as mentioned above used during the preparation processes.

2.4. Fabrication of Electrodes

100 mg of G, NG and x-NGM composites were mixed with 12.5 mg of carbon black in 2 mL of absolute ethanol, respectively. After ultrasonication for 10 min to make better dispersion, 0.5 g of ethyl cellulose and 1 mL of terpineol were added to the three kinds of suspensions. Ultrasonication for another 10 min was performed to obtain more even mixing. The subsequent stirring for 10 min was to evaporate some ethanol, to achieve an appropriate consistency of the G, NG and x-NGM slurries for fabricating the electrodes of ASSCs.

A polyimide (PI) tape with the dimension of 3.5 cm × 2.5 cm was attached and stuck to a graphite paper to obtain a PI/graphite flexible substrate. On the other hand, a square hole with the length of 1.6 cm was made in the center of transparency, which was placed upper the flexible substrate and fixed by the 3M tape. The transparency was closely attached to the substrate, and the effective area was thus defined. Afterward, an appropriate amount of the G, NG and x-NGM slurries was uniformly coated within the square on the flexible substrate by the doctor-blade method. After standing at the room temperature overnight, the transparency was removed to acquire the electrodes, which were then calcined at 200 °C for 1 h to eliminate organics. Three different mass loadings were used for coating active materials on the PI/graphite flexible substrates. The resulting electrodes were named as Gy, NGy, and x-NGMy, in which y was 1, 2, and 3, to represent the mass loadings of 1 mg, 2 mg, and 3 mg, respectively.

2.5. Characterization

The surface morphologies of active materials were examined by field-emission gun scanning electron microscopes (SEM) (Hitachi, Tokyo, Japan and JEOL, Tokyo, Japan). Their microstructures and lattice fringes were examined by a high-resolution transmission electron microscope (HRTEM) (JEOL, Tokyo, Japan). The elemental mappings were obtained by the energy-dispersive X-ray spectroscopy (EDS). The chemical compositions of active materials were examined by the X-ray photoelectron spectrometer (XPS) (Thermo VG-Scientific, Waltham, MA, USA). According to the binding energies of photoelectrons emitted from the surface, the chemical state of each element could be ascertained. The vibrational modes of molecules identified from the absorption characteristics by Raman spectroscopies (Horiba Jobin Yvon, Paris, France) and infrared (Bruker, Billerica, MA, USA) ranging from 400 cm^{-1} to 2000 cm^{-1} and 400 cm^{-1} to 4000 cm^{-1}, respectively, were used to determine the chemical compositions, bond configurations, and molecular structures of the active materials.

2.6. Electrochemical Measurements

The electrochemical properties were characterized by CV, galvanostatic charge/discharge (GCD), and electrochemical impedance spectroscopy (EIS), using a potentiostat/galvanostat (CH Instruments, Austin, TX, USA) as the analyzer. When the measurements on the electrodes coated with active materials were performed, a 5 M LiCl solution was employed as the electrolyte, and a three-electrode configuration consisting of a platinum (Pt) wire as the auxiliary electrode and silver chloride (Ag/AgCl) reference electrode was adopted.

The capacitance characteristics of the electrodes could be determined by the areas inside the CV curves obtained at different scan rates and the symmetry of the GCD curves obtained by different current densities. By Equations (1) and (2), the gravimetric specific capacitances C_{CV} and $C_{C\text{-}DC}$ of individual electrodes were calculated from the CV and GCD curves, respectively [60]:

$$C_{CV} = k\frac{\int i}{m \cdot s} \tag{1}$$

$$C_{C\text{-}DC} = k\frac{i \cdot \Delta t}{\Delta V \cdot m} \tag{2}$$

$$E_{EL}(Wh/kg) = (\frac{1}{4} \times C_{CV} \times V^2)/3.6 \tag{3}$$

$$E_{EL}(Wh/kg) = (\frac{1}{4} \times C_{C\text{-}DC} \times \Delta V^2)/3.6 \tag{4}$$

$$P_{EL}(W/kg) = E_{EL}/(\Delta t) \tag{5}$$

where k is the electrode constant (usually 2 for a single electrode and 4 for a couple of electrodes), i is the discharging current, $\int i$ is the integral area of a CV curve, m is the mass of electrode active materials, s is the scan rate (100 $mV \cdot s^{-1}$ in this work), Δt is the discharging time, and ΔV is the potential window subtracting the initial potential drop. In this study, the potential windows for individual electrodes were −1.9 V to 1.0 V. After substituting C_{CV} into Equation (3) and $C_{C\text{-}DC}$ into Equation (4), the energy densities of the electrodes (E_{EL}) could be calculated. By further substituting E_{EL} into Equation (5), the power densities of the electrodes (P_{EL}) were obtained [60]. On the other hand, the electronic and ionic transports across the interface of active material in the electrodes were investigated by EIS. The frequency range for EIS was 10^{-2} Hz to 10^5 Hz. The AC amplitude was set as 10 mV between two electrodes.

3. Results and Discussion

The surface morphologies of graphite oxide, G, NG, and x-NGM composites at different magnifications were examined by SEM. The ultrathin sheet-like structures of graphite oxide, G, and NG were observed, as displayed in Figure 1. There are fine needle structures in x-NGM, which grow on the surface of NG, as shown in Figure 1d–f. The changes in the content of Mn were discovered to impact the morphologies and structures of x-NGM composites significantly. When the concentration of $KMnO_4$ used during the preparation was higher to increase the content of Mn, a larger dimension of the needle structure resulted, as shown in Figure 1e,f. When the concentration of $KMnO_4$ was 26.7 mM to fabricate 3-NGM, the intertwined MnO_2 nanowires formed. When the $KMnO_4$ concentration increased to 35.6 mM and 44.5 mM to obtain 4-NGM and 5-NGM, the MnO_2 nanorods were observed, as displayed in Figure 1g,h. For further confirmation, elemental mappings were examined. Figure 2 demonstrates the presence of the C, O, N and Mn elements in 2-NGM and their even distributions. It verifies the successful preparation of the x-NGM composites as well. The sparser distribution of the N element was attributed to the relatively lower proportion of the N content in the composite.

Figure 3 displays the TEM micrographs and microstructures of NG, 2-NGM, and 3-NGM. The semitransparent membranous structure can be observed, as shown in Figure 3a,c,e, demonstrating the existence of NG in the composites. The lattice fringe spacing value of 0.34 nm corresponds to the d-spacing of G crystalline plane, indicating the successful formations of G and NG by the hydrothermal method [61]. Lots of fine needle structures appear on the NG surfaces of 2-NGM and 3-NGM, as shown in Figure 3c,e, respectively. The needle structure in 2-NGM is larger than that in 3-NGM. By the high-resolution atomic images in Figure 3d,f, the MnO_2 in the x-NGM composites is discovered to be a two-phase mixture. Namely, the co-existence of γ-MnO_2 and α-MnO_2. Another two lattice fringe spacing values of 0.212 nm and 0.239 nm correspond to the (200) plane of γ-MnO_2 and (211) plane of α-MnO_2, respectively [29], which also contribute to confirming the presence of MnO_2 and the successful preparation of the x-NGM composites.

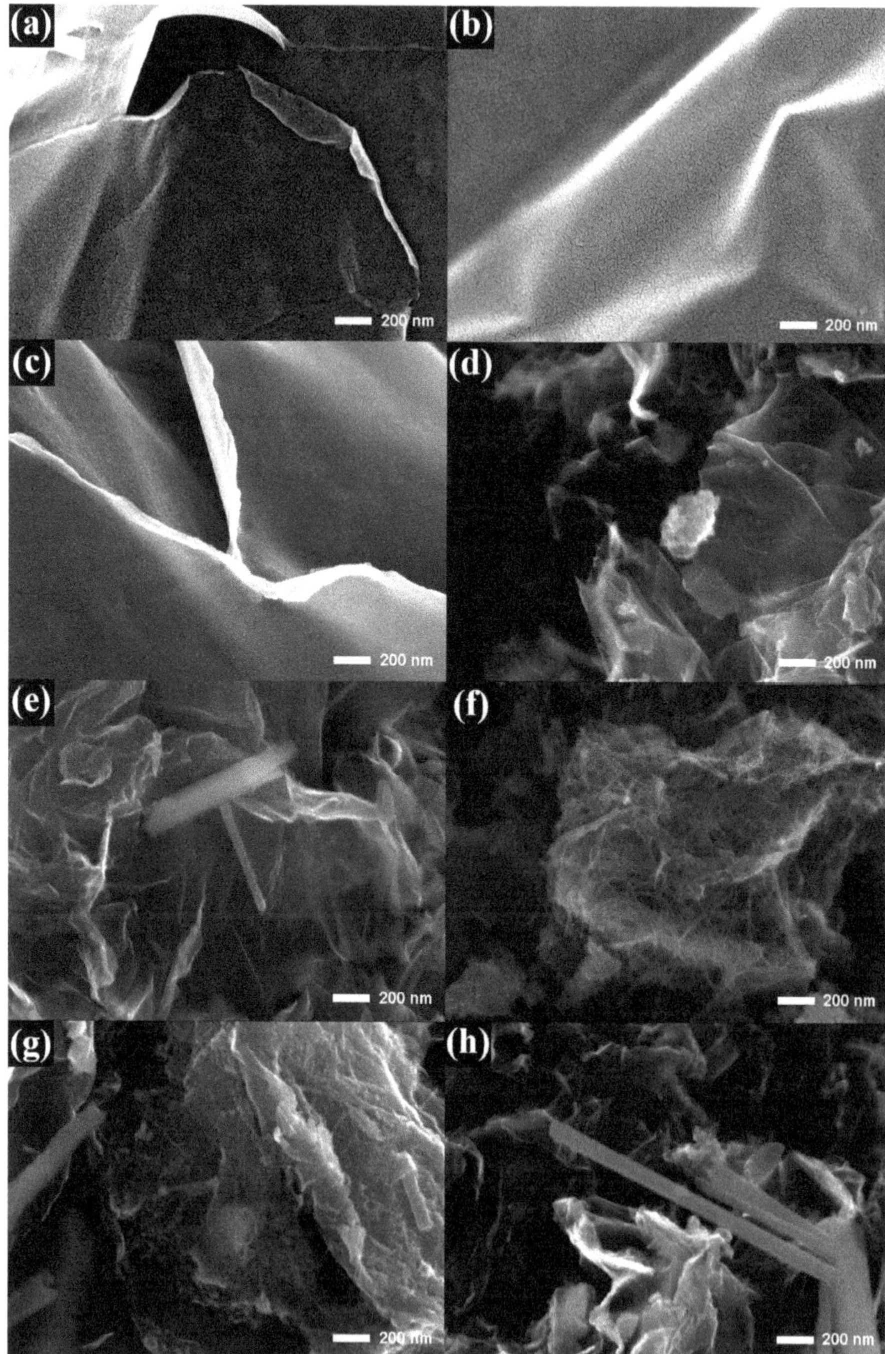

Figure 1. SEM micrographs of (**a**) graphite oxide, (**b**) G, (**c**) NG, (**d**) 1-NGM, (**e**) 2-NGM, (**f**) 3-NGM, (**g**) 4-NGM, and (**h**) 5-NGM.

The functional group types on the surface of a material can be identified by the FTIR technique. Figure 4 shows the FTIR spectra of graphite oxide, G, NG, and x-NGM composites. For graphite oxide, the main absorption peaks are at 1084 cm^{-1} and 1218 cm^{-1}, which are ascribed to the C-O stretchings. For G and NG, the two peaks at 1401 cm^{-1} and 1565 cm^{-1} are attributed to the O-H bending and C=C stretching, respectively. Graphite oxide, G, and NG all show the peak at 1720 cm^{-1}, which is ascribed to the C=O stretching. Another peak for G at 1214 cm^{-1} can be assigned to the C-O stretching [62–65]. For the N-containing composites (NG and x-NGM), the two main absorption peaks are at 1195 cm^{-1} and 1565 cm^{-1}, which are attributed to the C-N and C=N/C=C stretchings, respectively [66,67]. For the x-NGM composites, the two peaks at 438 cm^{-1} and 560 cm^{-1} can be ascribed to the Mn-O stretching [64–67]. One more peak for 2-NGM to 5-NGM is observed at 749 cm^{-1}, which can also be attributed to the Mn-O stretching, and not be found in the composite without or with too less content of Mn (such as 1-NGM). The FTIR results mentioned above also contribute to confirming the successful preparation of the G, NG, and x-NGM composites.

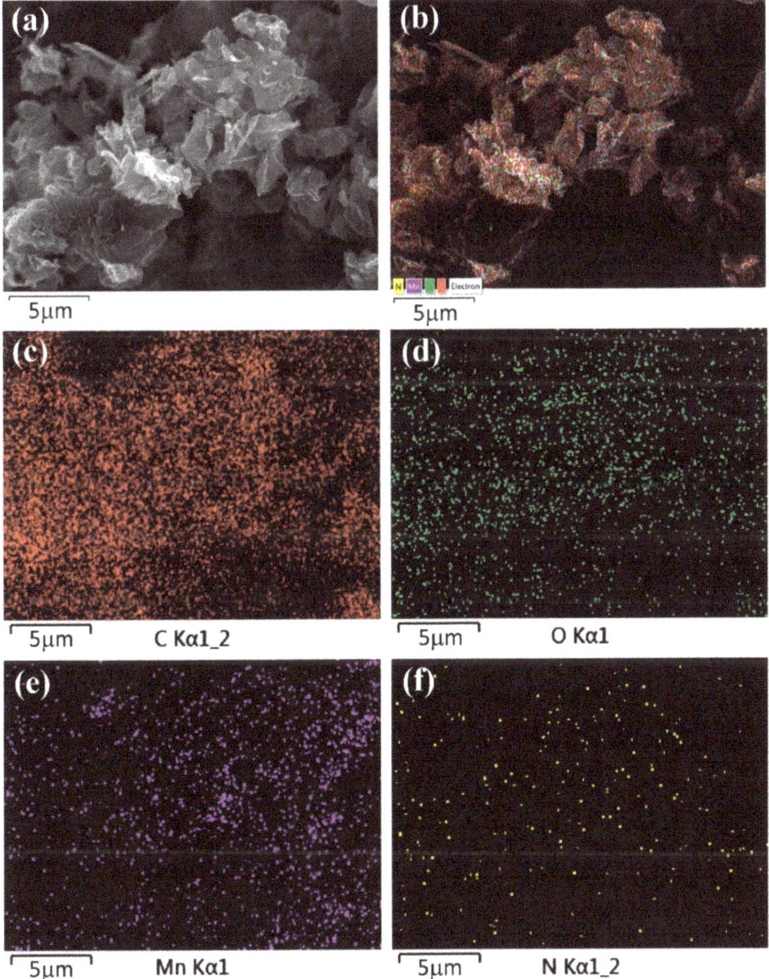

Figure 2. (**a**) SEM and (**b**) EDS layered images of 2-NGM. Elemental mappings of 2-NGM: (**c**) C, (**d**) O, (**e**) Mn, and (**f**) N.

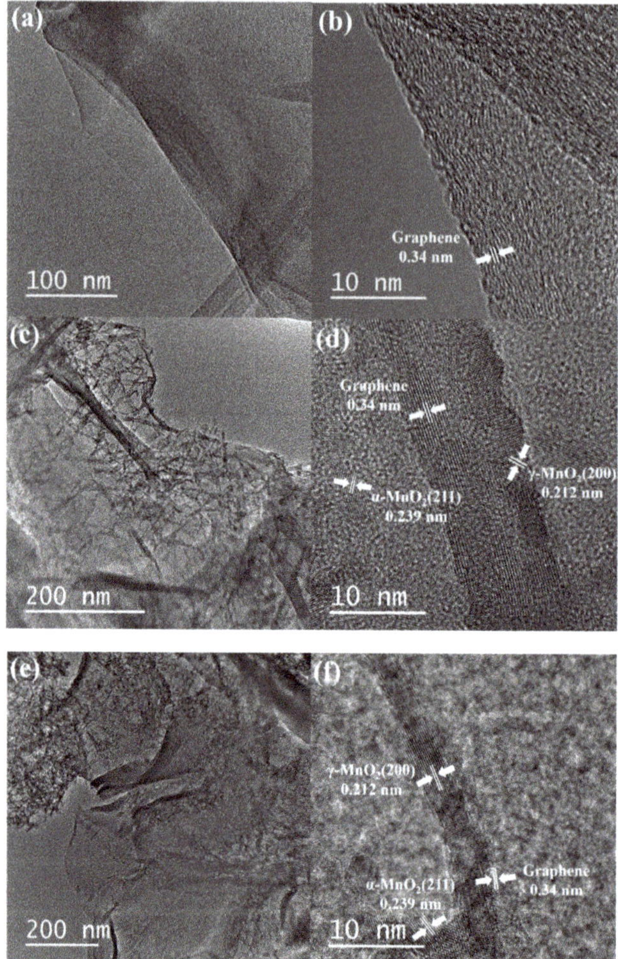

Figure 3. TEM micrographs and microstructures of (**a**,**b**) NG, (**c**,**d**) 2-NGM, and (**e**,**f**) 3-NGM.

Raman spectroscopy is a widely used technique to examine the structures and electronic properties of G and its derivatives. Figure 5 shows the Raman spectra of graphite oxide, G, NG, and x-NGM composites, ranging from 400 cm^{-1} to 2000 cm^{-1}. The two feature peaks at 1329 cm^{-1} and 1590 cm^{-1} are D and G bands, respectively [52,68,69]. The D band is due to the presence of disorders in sp^2-hybridized carbon systems. It can be used to estimate the defect level and content of impurity in the G sheets. The G band is derived from the stretching of sp^2-hybridized carbon-carbon bonds and highly sensitive to strain effects in the sp^2 system within the G sheets. Furthermore, the intensity ratio of D and G bands, I_D/I_G, can be considered as a measure of the relative concentration of local defects or interferences, i.e., can be used to estimate the extent of sp^3 graphite oxide converting to sp^2 G [68,69]. Thus, an increment of the I_D/I_G value implies an increase in the number of defects. From Figure 5a, the I_D/I_G value of graphite oxide obtained before the hydrothermal process is calculated to be 1.73. However, those of the G, NG, and x-NGM composites drop to in between 1.55 to 1.69 after the hydrothermal process. It can be then deduced from the reduced I_D/I_G values that the hydrothermal process could remove oxygen-containing functional groups and reduce graphite oxide to G successfully. This again helps to ascertain the presence of G in NG and x-NGM composites. Since the incident light wavelength of the Raman spectrometer influences excitation efficiency, the wavelength of 532 nm regarded as comparatively of less negative impact on enhancing the characteristic peak of Mn-O bonds was chosen to excite the x-NGM composites. However, as shown in Figure 5a, it still unavoidably resulted in weaker scattering intensity and thereby less distinct Mn-O characteristic peaks. After magnification, the peak at around 560 cm^{-1} attributed to the stretching vibration of Mn-O bonds [52] can be more clearly seen in Figure 5b, again confirming that the hydrothermal preparation of x-NGM composites was successful.

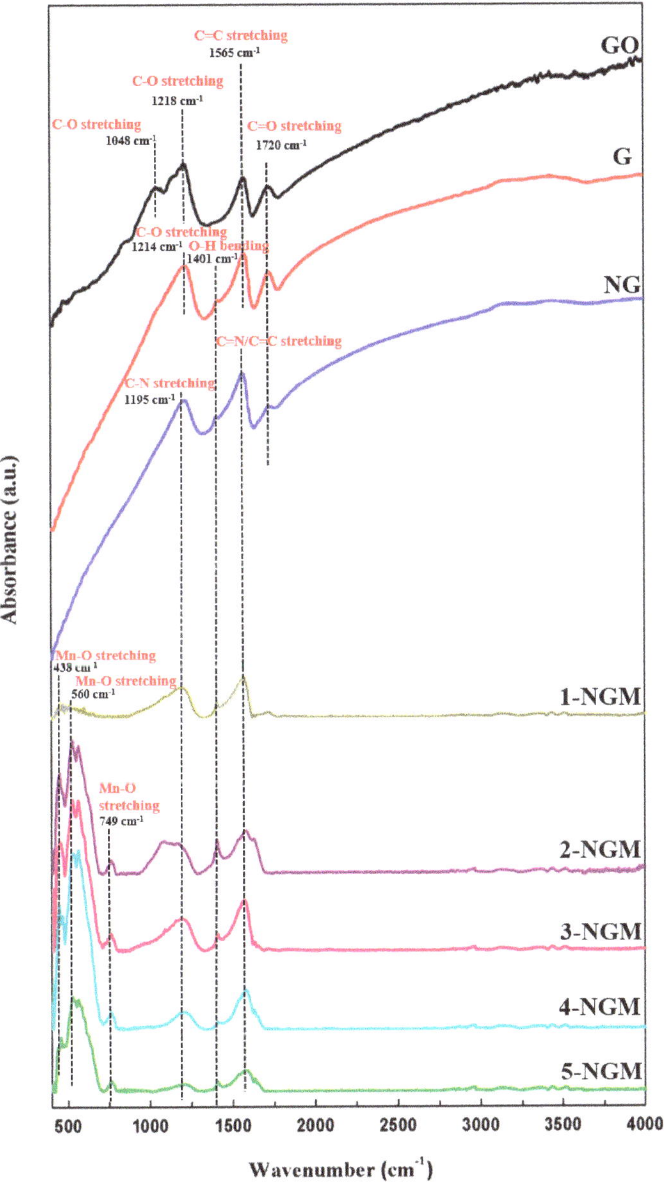

Figure 4. FTIR spectra of graphite oxide, G, NG, and x-NGM composites.

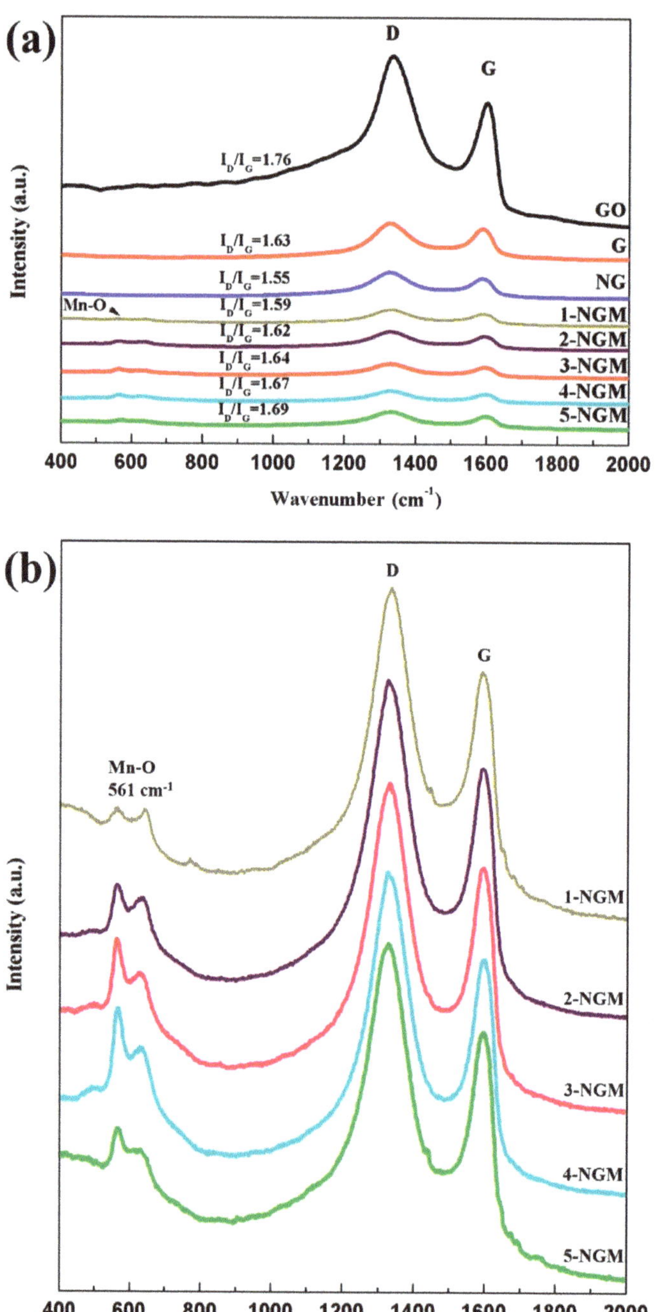

Figure 5. (**a**) Raman spectra of graphite oxide, G, NG, and x-NGM composites; (**b**) enlarged Raman spectra of x-NGM composites.

Figure 6 shows the XPS spectra of graphite oxide, G, NG, and x-NGM composites. The composition of a composite and chemical states of elements can be investigated by XPS. Figure 6a displays the C 1s spectra. For graphite oxide and G, they have a strong energy peak centered at approximately 283.8 eV, which is assigned to the C=C bonds (sp^2-hybridized carbon atoms). Another weak peak with higher binding energy at 285.6 eV can be assigned to the C-O bonds (oxygenated carbon atoms) [62,63,70]. There is a weaker peak for graphite oxide at 287.4 eV, which is assigned to the C=O bonds. G shows an even weaker peak at 288.6 eV, which is assigned to the O-C=O bonds [52,62,63,70]. For NG and x-NGM composites, they also have energy peaks ascribed to the C=C bonds (symbol of the presence of G), centered at approximately 284.0 eV. When the containing of Mn in x-NGM to diminish the C=C bonds and sp^2-hybridized carbon atoms, the peak intensity is found to be weakened. Another two weaker peaks of NG and x-NGM, centered at approximately 285.3 eV and 286.6 eV, can be assigned to the C=N and C-N bonds, respectively [62,63,69], which stand for another evidence for the existence of N element in the composites.

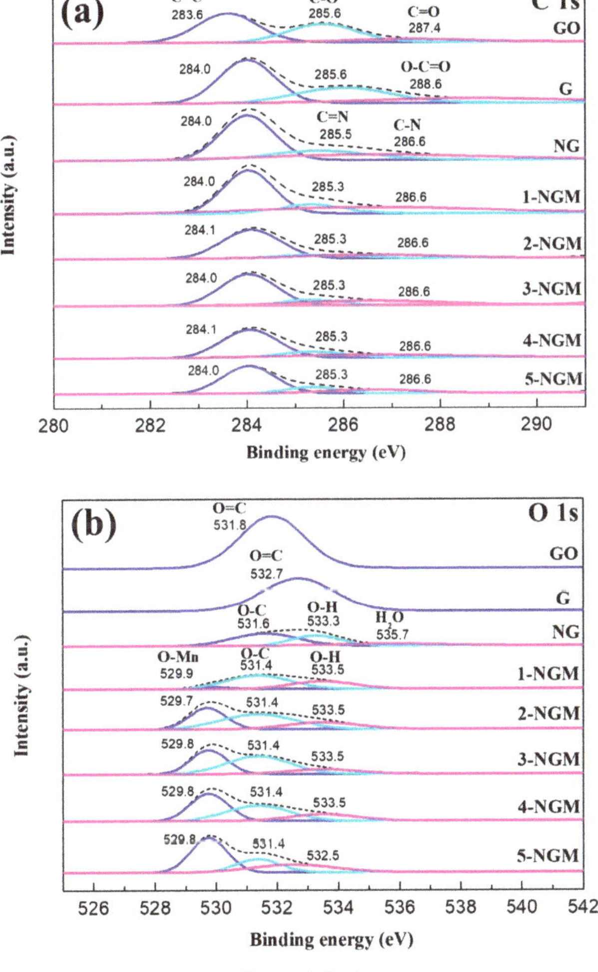

Figure 6. *Cont.*

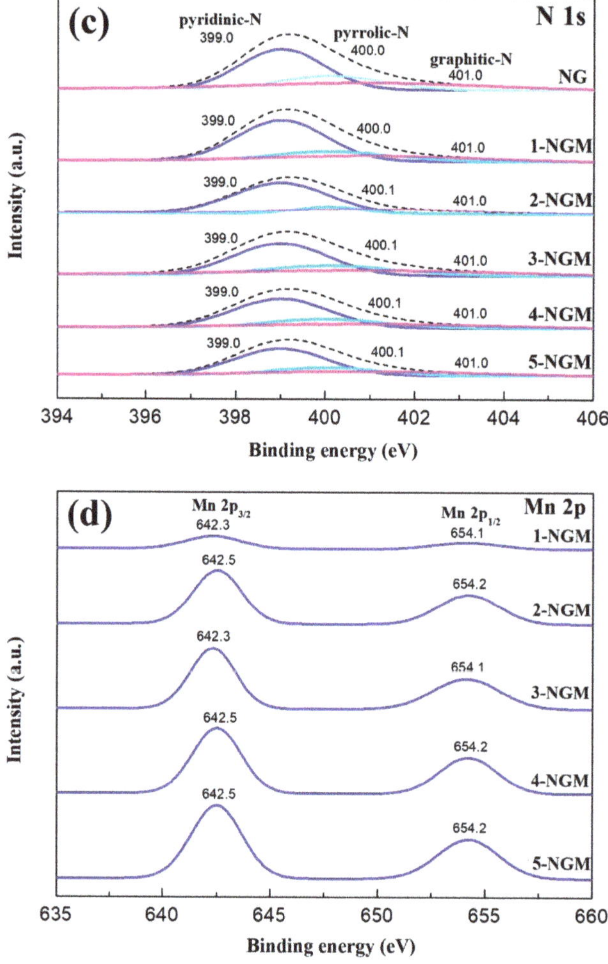

Figure 6. XPS spectra of graphite oxide, G, NG, and x-NGM composites: (**a**) C 1s, (**b**) O 1s, (**c**) N 1s, and (**d**) Mn 2p.

Figure 6b displays the O 1s spectra. Both graphite oxide and G show an energy peak ascribed to the C=O bonds at 531.8 eV and 532.7 eV, respectively. The three energy peaks of NG at 531.6 eV, 533.3 eV, and 535.7 eV, can be assigned to the bondings of O-C, O-H, and H_2O, respectively [52,62,63]. The five x-NGM composites have similar O 1s spectra. They all exhibit the three energy peaks centered at approximately 529.8 eV, 531.4 eV, and 533.5 eV, which correspond to the Mn-O, C-O, and O-H bonds, respectively [52,62,63,69]. The increase in the content of Mn is found to enhance the intensity of the 529.8 eV energy peak. Figure 6c shows the N 1s spectra. All the six NG and x-NGM composites exhibit a strong peak at 399.0 eV and two weak peaks at 400.1 eV and 401.0 eV, which can be attributed to the three types of N-containing species on the surface: pyridinic-N, pyrrolic-N, and graphitic-N (quaternary N), respectively [52,70,71]. Again, the presence of the N element in the six active materials is confirmed. The fabrication of NG from G is demonstrated to be successful as well. Figure 6d shows the high-resolution Mn 2p spectra. All the five x-NGM composites exhibit the two peaks centered at approximately 642.3 eV and 654.1 eV, which are ascribed to the Mn $2p_{3/2}$ and Mn $2p_{1/2}$ spin-orbit splitting states, respectively. The separation of spin energy between the two peaks is 11.8 eV, indicating the oxidation state of Mn is +4. [52]. Moreover, the intensity of the two peaks is found to increase with an increased content of Mn. This not only confirms the presence of Mn in the x-NGM composites, but also demonstrates that the use of a hydrothermal method for preparation of the composites containing both NG and MnO_2 was successful [52,62,63,72]. The XPS surveys have confirmed the presence of C, Mn, O, and N on the surface of the x-NGM composites.

EIS is a technique used for acquiring information of internal impedances in an electrochemical system. The electronic and ionic transports along the bulk and across the interface of active material in the electrode were thus investigated. The Nyquist plots of the 21 electrodes with different active materials are displayed in Figure 7, where the compressed semicircles at the high and medium frequency regions are related to the electronic transport resistance, a kinetic-controlled process. The line tail connecting the semicircle at the low-frequency region is associated with the ionic diffusion resistance, a thermodynamic-controlled process. The equivalent circuit for the EIS analysis is also depicted in Figure 7, which includes [73]: (1) the charge transfer impedance in the high-frequency region at the electrode/electrolyte interface (R_{CT}), (2) the solution resistance (R_S), which is the contact series resistance between the substrate and current collector, (3) Warburg impedance (W), which is the diffusion resistance of ions in the electrolyte and in relation to the slope of the line tail in the low-frequency region, (4) the electric double-layer capacitor (C_1) [65,74–76]. The corresponding R_{CT} values obtained by simulation are listed in Table 1. By contrast, G1, NG1, 1-NGM1, 2-NGM1, 3-NGM1, 4-NGM1, and 5-NGM1 are found to exhibit smaller semicircles in the high-frequency region. Their R_{CT} values are 3.27 Ω, 2.22 Ω, 2.17 Ω, 1.28 Ω, 1.15 Ω, 8.06 Ω, and 9.29 Ω, respectively. When the mass loading of active materials on the substrate increases to 2 mg, the R_{CT} values of G2, NG2, 1-NGM2, 2-NGM2, 3-NGM2, 4-NGM2, 5-NGM2 are 3.52 Ω, 2.98 Ω, 2.49 Ω, 1.35 Ω, 1.23 Ω, 8.43 Ω, and 9.40 Ω, respectively. When the mass loading further increases to 3 mg, the R_{CT} values of G3, NG3, 1-NGM3, 2-NGM3, 3-NGM3, 4-NGM3, and 5-NGM3 are 12.83 Ω, 9.21 Ω, 8.42 Ω, 2.14 Ω, 1.70 Ω, 9.46 Ω, and 11.60 Ω, respectively. According to the significantly increased R_{CT} and larger semicircle in the high-frequency region, it can be then concluded that the preferred mass loading of active materials on the PI/graphite flexible substrate is 1 mg.

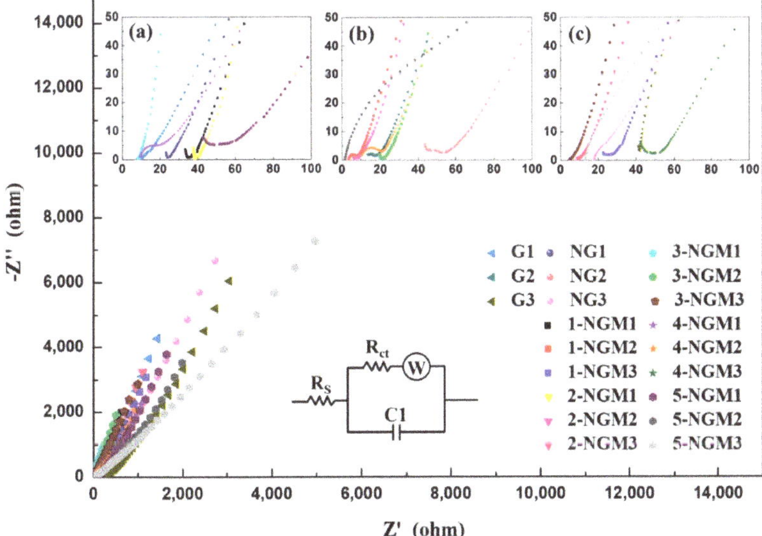

Figure 7. Nyquist plots of the 21 electrodes with Gy, NGy, and x-NGMy composites. Insets are enlargements when the mass loading is (**a**) 1 mg, (**b**) 2 mg, and (**c**) 3 mg.

Table 1. R_{CT} values of the 21 electrodes with Gy, NGy, and x-NGMy composites obtained by EIS simulation.

Electrode	R_{CT} (Ω)	Electrode	R_{CT} (Ω)	Electrode	R_{CT} (Ω)
G1	3.27	1-NGM2	2.49	3-NGM3	1.70
G2	3.52	1-NGM3	8.42	4-NGM1	8.06
G3	12.83	2-NGM1	1.28	4-NGM2	8.43
NG1	2.22	2-NGM2	1.35	4-NGM3	9.46
NG2	2.98	2-NGM3	2.14	5-NGM1	9.29
NG3	9.21	3-NGM1	1.15	5-NGM2	9.40
1-NGM1	2.17	3-NGM2	1.23	5-NGM3	11.60

As shown in Table 1, after N was involved in the active materials to obtain NG composites, the R_{CT} value decreased from 3.27 Ω (G1) to 2.22 Ω (NG1). The interconnected NG component can effectively facilitate electronic transport. When N and MnO_2 were simultaneously involved in the x-NGM composites, the R_{CT} value was further reduced from 2.22 Ω (NG1) to 1.15 Ω (3-NGM1). This indicates that the co-existence of NG and MnO_2 in an active material could even more improve charge transfer. The diffusion/transport properties of electrolyte ions to the electrode surface can be examined by the linear response (slope) of the Nyquist plot in the low-frequency region. An increased slope usually illustrates a lower diffusion resistance, faster ionic transport, and thereby enhanced capacitive property [74,77]. As displayed in Figure 7, compared to those of G1 and NG1 electrodes, the Nyquist plot of the 1-NGM1 electrode has a larger slope in the low-frequency region, indicating its better ionic diffusion and higher conductivity since it contains NG and MnO_2 simultaneously. The impact of the Mn content on R_{CT} was further investigated. Among the 15 electrodes with x-NGM active materials, the R_{CT} values for 1-NGM1, 2-NGM1, 3-NGM1, 4-NGM1, and 5-NGM1 are 2.17 Ω, 1.28 Ω, 1.15 Ω, 8.06 Ω, and 9.29 Ω, respectively. The 3-NGM1 electrode exhibits the smallest R_{CT}, indicating its best charge transfer efficiency. Its Nyquist plot in the low-frequency region is almost vertical and shows the largest slope, representing the best ionic diffusion and capacitive properties of the 3-NGM1 electrode. It can be then deduced that the MnO_2 nanowires growing on the NG surface provide more effective contacts between the electrode and electrolyte ions, and charge transfer is thus improved. However, excess Mn in the 4-NGM1 and 5-NGM1 electrodes causes increased R_{CT} values due to fewer contacts. Ion diffusion and charge transfer capacity are thereby suppressed, and a smaller slope of the Nyquist plot in the low-frequency region results. By the aforementioned results, it is confirmed that both the mass loading and content of Mn in an active material electrode affect conductivity. The best charge transfer efficiency can be obtained only when the mass loading is 1 mg and the content of Mn in x-NGM composites is optimized (x = 3).

The capacitive characteristic of an active material electrode can be evaluated by integrating the area inside a CV curve loop. Figure 8a,b shows the CV curves of the 21 electrodes with Gy, NGy, and x-NGMy composites, obtained at the scanning rate of 100 $mV \cdot s^{-1}$ with a fixed potential range of −2.9 V to 1.0 V. All x-NGMy composites show redox waves, indicating that the faradaic phenomena occurred during the charge/discharge process. No redox peaks are found for Gy and NGy composites. Their CV curves exhibit similar rectangular and symmetrical form, which is related to the characteristics of an electric double-layer capacitor. In Figure 8a, the CV curve loops of the 1-NGMy electrodes are the largest, indicating that they have the highest specific capacitances, followed by the NGy and then the Gy electrodes. Among them, the 1-NGM1 electrode has the largest area inside the loop, giving rise to a high specific capacitance of 416 $F \cdot g^{-1}$. Its energy and power densities obtained by calculations are 243.2 $Wh \cdot kg^{-1}$ and 959.9 $W \cdot kg^{-1}$, respectively. The capacitance enhancement can be ascribed to the synergistic effect of the higher conductivity by NG and the larger specific surface area by MnO_2 nanostructures. Afterward, the content of Mn was tuned by using various $KMnO_4$ concentrations. The impacts of mass loading on the capacitive property were also studied by using 1 mg, 2 mg, and 3 mg of active materials (y = 1, 2, and 3) for each x. From the CV curve loops of the x-NGMy electrodes in Figure 8b, the 3-NGM1 electrode shows the best capacitance performance with the highest specific capacitance of 638 F·g^{-1}. Its energy and power densities are 372.7 Wh·kg^{-1} and 4731.1 W·kg^{-1}, respectively. All the capacitance parameters acquired by calculations are listed in Table 2, which reveals that there was a most appropriate $KMnO_4$ concentration, i.e., 26.72 mM when x = 3, to achieve an NGM composite with the optimum Mn content. Moreover, the increase of specific capacitance is more significant by pseudocapacitive MnO_2 than NG.

Table 2. Capacitance parameters obtained from the CV results of the 21 electrodes with Gy, NGy, and x-NGMy composites.

Electrode	Scan Rate (mV·s^{-1})	Specific Capacitance (F·g^{-1})	Energy Density (Wh·kg^{-1})	Power Density (W·kg^{-1})
G1	100	243	141.9	930.5
G2	100	99	57.5	889.0
G3	100	14	8.3	381.8
NG1	100	247	144.4	810.9
NG2	100	104	60.6	375.5
NG3	100	45	26.1	224.9
1-NGM1	100	416	243.2	959.9
1-NGM2	100	223	130.4	662.3
1-NGM3	100	154	90.1	572.9
2-NGM1	100	516	301.6	6556.8
2-NGM2	100	267	155.7	6370.1
2-NGM3	100	173	100.7	6407.3
3-NGM1	100	638	372.7	4731.1
3-NGM2	100	298	174.2	7208.0
3-NGM3	100	192	112.0	6001.7
4-NGM1	100	141	82.0	7959.5
4-NGM2	100	62	36.4	6689.0
4-NGM3	100	36	20.9	6105.0
5-NGM1	100	94	55.0	7799.2
5-NGM2	100	40	23.4	4918.4
5-NGM3	100	12	7.1	2227.0

By the interlaced nanowire structures of MnO_2, the contact between the electrode surface and electrolyte ions and the use of active materials are both improved. There are more sites for electrochemical reactions to occur, and enhanced diffusion of ions in the electrolyte favorable for capacitive characteristic has resulted. The higher $KMnO_4$ concentrations during the preparation process, 35.62 mM and 44.53 mM when x = 4 and 5, led to excess MnO_2 in the NGM composites, which grew into nanorod structures detrimental to more contact between the electrode surface and electrolyte ions. Inferior ion diffusion and worse capacitive characteristics of the 4-NGMy and 5-NGMy electrodes are thereby caused, as shown in Figure 8b and Table 2. It can be also seen from Table 2 that the mass loading of active material on the flexible electrode is also critical. 1 mg has been demonstrated to be the most appropriate mass loading. 2 mg and 3 mg cause overloaded active materials and thus reduced specific capacitance, energy, and power densities. By plotting energy density vs. power density obtained from the CV results and calculated by Equations (3) and (5), a Ragone plot is achieved, as shown in Figure 8c, which again demonstrates the best electrochemical performance of 3-NGM1 among the 21 x-NGMy electrodes when x = 3 and the mass loading of 1 mg were used.

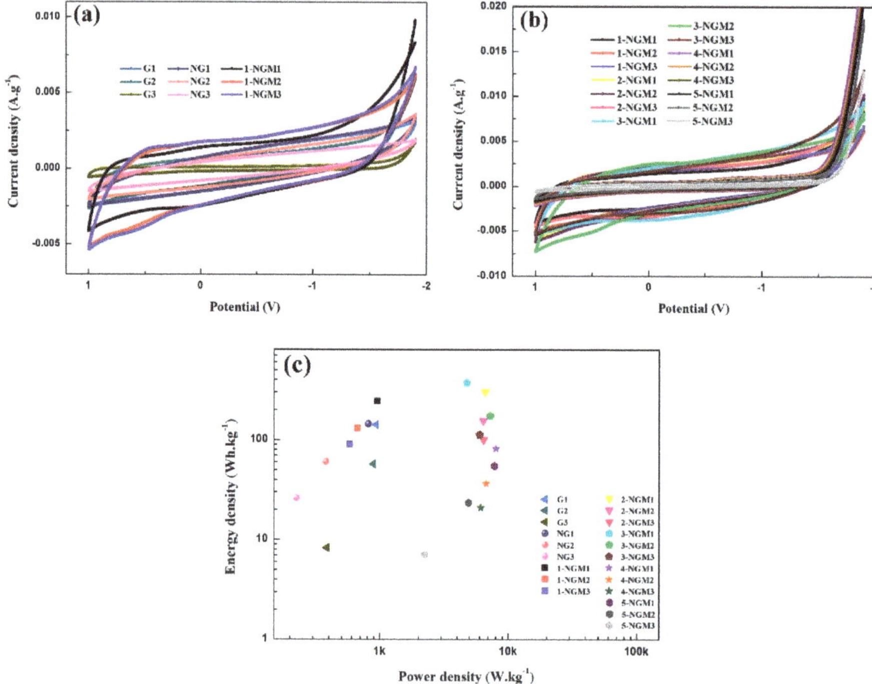

Figure 8. CV curves of the 21 electrodes with (**a**) Gy, NGy, 1-NGMy, and (**b**) x-NGMy composites. (**c**) Ragone plot obtained from the CV results.

The symmetry of charge and discharge curves can be investigated to understand capacitive behavior. Figure 9a shows the GCD curves of the 9 electrodes with Gy, NGy, and 1-NGMy composites, obtained by a fixed potential range of -1.2 V to 1.0 V under different current densities. Those of the 1-NGMy electrodes are slightly distorted from the ideal triangle shape because of the pseudocapacitive contribution from MnO_2. The curvature implies that they are typical Faraday capacitance curves [78]. It is revealed that the 1-NGMy electrodes have the best capacitive characteristics, followed by the NGy electrodes, and the Gy electrodes are the worst. The 1-NGM1 electrode exhibits a longer charge and discharge time at a current density of 0.2 $A{\cdot}g^{-1}$, resulting in a specific capacitance of 188 $F{\cdot}g^{-1}$, and the corresponding energy and power densities are 63.1 $Wh{\cdot}kg^{-1}$ and 249.2 $W{\cdot}kg^{-1}$, respectively. Afterward, the impacts of the Mn content and mass loading on the capacitive parameters of the electrodes were also explored. The specific capacitances can be calculated from the GCD curves by Equation (2), as listed in Table 3. Figure 9b shows the plots of specific capacitance vs. current density for the 21 electrodes. The Gy, NGy, 1-NGMy, 4-NGMy, and 5-NGMy electrodes cannot endure the current densities larger than 1 $A{\cdot}g^{-1}$. The 2-NGM2 electrode can endure only the current densities of 1 $A{\cdot}g^{-1}$, 3 $A{\cdot}g^{-1}$, and 5 $A{\cdot}g^{-1}$. The 3-NGM2 electrode can endure the current densities of 1 $A{\cdot}g^{-1}$, 3 $A{\cdot}g^{-1}$, 5 $A{\cdot}g^{-1}$, and 7 $A{\cdot}g^{-1}$. The 2-NGM3 and 3-NGM3 electrodes can endure only the current densities of 1 $A{\cdot}g^{-1}$ and 3 $A{\cdot}g^{-1}$, whereas the 2-NGM1 and 3-NGM1 electrodes can endure all current densities. Among the 21 electrodes, the 3-NGM1 electrode exhibits the best endurance. Its energy and power densities are 86.7 $Wh{\cdot}kg^{-1}$ and 1100.0 $W{\cdot}kg^{-1}$, respectively. The above results have also confirmed the optimum conditions for the mass loading and content of Mn.

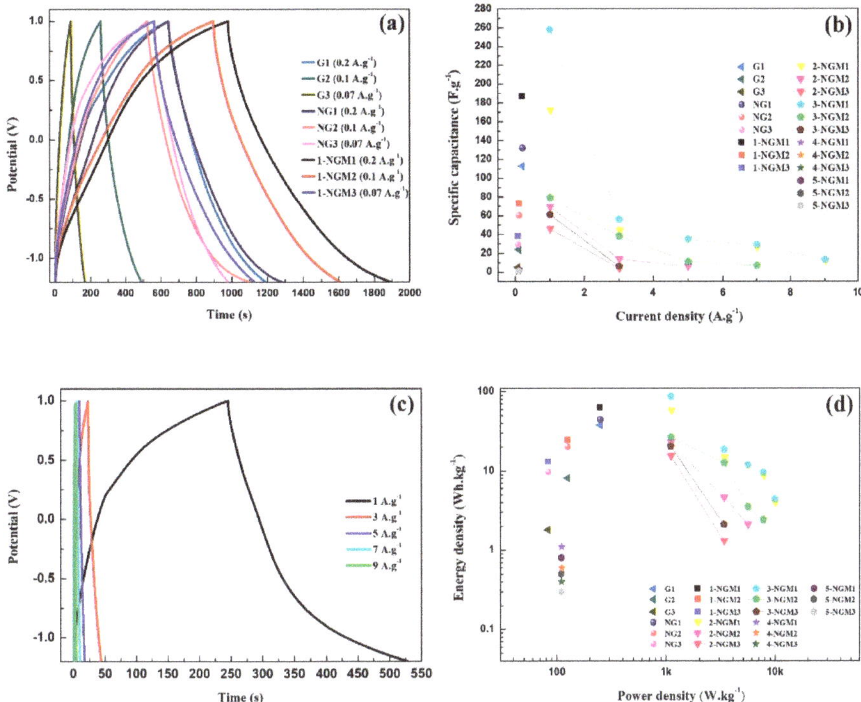

Figure 9. (**a**) GCD curves of the 9 electrodes with Gy, NGy, and 1-NGMy composites. (**b**) Plots of specific capacitance vs. current density for the 21 electrodes with Gy, NGy, and x-NGMy composites. (**c**) GCD curves of the 3-NGM1 electrode under the current densities of 1 A·g^{-1} to 9 A·g^{-1}. (**d**) Ragone plot obtained from the GCD results.

Since the 3-NGM1 electrode has shown the best sustainable ability to permit its higher charging capacity, it was selected to perform further GCD investigation by different current densities, are shown in Figure 9c. The rapid intercalation/deintercalation of metallic cations in an active material reveals the redox concerning the oxidation state transitions between Mn (III) and Mn (IV). It is inferred that at higher current densities, only the external surface of an active material is involved in charge/discharge, leading to insufficient redox and relatively lower specific capacitances. The charge/discharge time decreases along with a small number of electrolyte ions occupying the active sites. By contrast, at lower current densities, more internal and external active sites are involved, attaining more complete redox reactions and higher specific capacitances [79]. The increased charge/discharge time results from most electrolyte ions being anchored to the active sites at the interface. As displayed in Table 3, when the current density applied to the 3-NGM1 electrode increases from 1 A·g^{-1} to 9 A·g^{-1}, the specific capacitance reduces from 258 F·g^{-1} to 13 F·g^{-1}. The massive capacitance decay implies that the rate capability of the x-NGMy electrodes still needs considerable improvement. Moreover, it is perceived from Table 3 that the mass loading of active materials on the flexible electrode is very critical. 1 mg has been proven to be the optimum. Both 2 mg and 3 mg are overloads to cause lower conductivity and inferior capacitive parameters. For the Gy, NGy, and x-NGMy electrodes, their energy and power densities calculated from the GCD results are plotted as a Ragone plot, as shown in Figure 9d. Consistent with the EIS and CV results, the synergistic effect of NG with MnO_2 is demonstrated again, to promote reversible redox reactions on the pseudocapacitive materials and play great impacts on the capacitance characteristics of the electrodes.

Table 3. Capacitance parameters obtained from the GCD results of the 21 electrodes with Gy, NGy, and x-NGMy composites by different current densities.

Electrode	Current Density ($A\cdot g^{-1}$)	Specific Capacitance ($F\cdot g^{-1}$)	Energy Density ($Wh\cdot kg^{-1}$)	Power Density ($W\cdot kg^{-1}$)
G1	0.2	113	38.0	249.2
G2	0.1	24	8.1	124.6
G3	0.07	5	1.8	83.1
NG1	0.2	132	44.4	249.2
NG2	0.1	60	20.1	124.6
NG3	0.07	29	9.7	83.1
1-NGM1	0.2	188	63.1	249.2
1-NGM2	0.1	73	24.5	124.6
1-NGM3	0.07	39	13.1	83.1
2-NGM1	1	172	57.8	1100.0
2-NGM2	1	69	23.1	1100.0
2-NGM3	1	46	15.4	1100.0
3-NGM1	1	258	86.7	1100.0
3-NGM2	1	79	26.6	1100.0
3-NGM3	1	61	20.5	1100.0
2-NGM1	3	44	14.7	1100.0
2-NGM2	3	14	4.6	3300.0
2-NGM3	3	4	1.3	3300.0
3-NGM1	3	56	18.8	3300.0
3-NGM2	3	38	12.7	3300.0
3-NGM3	3	6	2.1	3300.0
2-NGM1	5	35	11.8	5500.0
2-NGM2	5	6	2.1	5500.0
3-NGM1	5	36	11.9	5500.0
3-NGM2	5	11	3.5	5500.0
2-NGM1	7	26	8.6	7700.0
3-NGM1	7	29	9.6	7700.0
3-NGM2	7	7	2.4	7700.0
2-NGM1	9	12	3.9	9900.0
3-NGM1	9	13	4.4	9900.0
4-NGM1	0.1	3	1.1	110.0
4-NGM2	0.1	2	0.6	110.0
4-NGM3	0.1	1	0.4	110.0
5-NGM1	0.1	2	0.8	110.0
5-NGM2	0.1	1.6	0.5	110.0
5-NGM3	0.1	1	0.3	110.0

4. Conclusions

In this study, x-NGM composites consisting of NG and MnO_2 with various Mn contents were fabricated by a low-cost hydrothermal method. By SEM, TEM, EDS mappings, XPS, FTIR and Raman spectra, the presence of NG and MnO_2 was confirmed, and the successful preparation of the composites was demonstrated. The microstructure analysis by TEM manifested that the MnO_2 in the x-NGM composites was a two-phase mixture of γ- and α-MnO_2. According to the EIS results, the NG component was found to reduce R_{CT} effectively due to its good conductivity. The co-existence of NG and MnO_2 in an active material led to a more reduced R_{CT} and further improved charge transfer. Among the 15 electrodes with x-NGM active materials, the 3-NGM1 electrode exhibited the smallest R_{CT}, indicating its best charge transfer efficiency. Its Nyquist plot in the low-frequency region had the largest slope, implying a lower diffusion impedance, more rapid ionic diffusion, and enhanced capacitive property. Both the mass loading and content of Mn in an active material electrode were crucial. The best electrochemical performance was achieved when the mass loading of active materials on the PI/graphite flexible substrate was 1 mg and x = 3 to obtain the optimized Mn content in the x-NGM composites. Excess Mn caused decreased contacts between the electrode and electrolyte ions, leading to increased R_{CT}, and suppressed ionic diffusion. Among the 21 electrodes with Gy, NGy, and x-NGMy composites, the 3-NGM1 electrode exhibited the best sustainable ability. However, its rate capability still required large improvement. After calculation of the CV results, it showed a high specific capacitance of 638 F·g^{-1}, and the corresponding energy and power densities were 372.7 Wh·kg^{-1} and 4731.1 W·kg^{-1}, respectively. The enhancement was ascribed to the synergistic effect of the higher conductivity by NG and the larger specific surface area by MnO_2 nanostructures. Moreover, the increase of specific capacitance was found to be more significant by the pseudocapacitive MnO_2 than NG.

Author Contributions: Methodology, C.-P.C.; Validation, H.-Y.C.; Data Curation, H.-Y.C.; Investigation, H.-Y.C.; Formal Analysis, H.-Y.C.; Writing-Original Draft Preparation, C.-P.C.; Writing-Review & Editing, C.-P.C.; Visualization, C.-P.C.; Supervision, C.-P.C.; Project Administration, C.-P.C.; Funding Acquisition, C.-P.C.

Funding: This research and the APC were funded by the Ministry of Science and Technology, Taiwan under the grant number of MOST 107-2221-E-260-002-.

Acknowledgments: Supports from the Ministry of Science and Technology Taiwan and National Chi Nan University are gratefully appreciated.

Conflicts of Interest: The authors declare no conflicts of interest.

References

1. Simon, P.; Gogotsi, Y. Materials for electrochemical capacitors. *Nat. Mater.* **2008**, *7*, 845–854. [CrossRef] [PubMed]
2. Merlet, C.; Rotenberg, B.; Madden, P.A.; Taberna, P.L.; Simon, P.; Gogotsi, Y.; Salanne, M. On the molecular origin of supercapacitance in nanoporous carbon electrodes. *Nat. Mater.* **2012**, *11*, 306–310. [CrossRef] [PubMed]
3. Kyeremateng, N.A.; Brousse, T.; Pech, D. Microsupercapacitors as miniaturized energy-storage components for on-chip electronics. *Nat. Nanotechnol.* **2017**, *12*, 7–15. [CrossRef] [PubMed]
4. Stauss, S.; Honma, I. Biocompatible batteries-materials and chemistry, fabrication, applications, and future prospects. *Bull. Chem. Soc. Jpn.* **2018**, *91*, 492–505. [CrossRef]
5. Pedico, A.; Lamberti, A.; Gigot, A.; Fontana, M.; Bella, F.; Rivolo, P.; Cocuzza, M.; Pirri, C.F. High-performing and stable wearable supercapacitor exploiting rGO aerogel decorated with copper and molybdenum sulfides on carbon fibers. *ACS Appl. Energy Mater.* **2018**, *1*, 4440–4447. [CrossRef]
6. Abdelkader, A.; Rabeh, A.; Mohamed, A.D.; Mohamed, J. Multi-objective genetic algorithm based sizing optimization of a stand-alone wind/PV power supply system with enhanced battery/supercapacitor hybrid energy storage. *Energy* **2018**, *163*, 351–363. [CrossRef]
7. Scalia, A.; Bella, F.; Lamberti, A.; Bianco, S.; Gerbaldi, C.; Tresso, E.; Pirri, C.F. A flexible and portable powerpack by solid-state supercapacitor and dye-sensitized solar cell integration. *J. Power Sources* **2017**, *359*, 311–321. [CrossRef]
8. Chen, H.Y.; Zeng, S.; Chen, M.H.; Zhang, Y.Y.; Zheng, L.X.; Li, Q.W. Oxygen evolution assisted fabrication of highly loaded carbon nanotube/MnO_2 hybrid films for high-performance flexible pseudosupercapacitors. *Small* **2016**, *12*, 2035–2045. [CrossRef] [PubMed]
9. Wang, J.G.; Kang, F.Y.; Wei, B.Q. Engineering of MnO_2-based nanocomposites for high-performance supercapacitors. *Prog. Mater. Sci.* **2015**, *74*, 51–124. [CrossRef]
10. Wang, J.G.; Liu, H.Y.; Liu, H.Z.; Fu, Z.H.; Nan, D. Facile synthesis of microsized MnO/C composites with high tap density as high performance anodes for Li-ion batteries. *Chem. Eng. J.* **2017**, *328*, 591–598. [CrossRef]
11. Zhu, Y.Q.; Cao, C.B.; Tao, S.; Chu, W.S.; Wu, Z.Y.; Li, Y.D. Ultrathin nickel hydroxide and oxide nanosheets: Synthesis, characterizations and excellent supercapacitor performances. *Sci. Rep.* **2015**, *4*, 5787. [CrossRef] [PubMed]
12. Zhang, L.L.; Zhao, X.S. Carbon-based materials as supercapacitor electrodes. *Chem. Soc. Rev.* **2009**, *38*, 2520–2531. [CrossRef] [PubMed]
13. Wu, X.L.; Wen, T.; Guo, H.L.; Yang, S.B.; Wang, X.K.; Xu, A.W. Biomass-derived sponge-like carbonaceous hydrogels and aerogels for supercapacitors. *ACS Nano* **2013**, *7*, 3589–3597. [CrossRef] [PubMed]

14. Khan, A.H.; Ghosh, S.; Pradhan, B.; Dalui, A.; Shrestha, L.K.; Acharya, S.; Ariga, K. Two-dimensional (2D) nanomaterials towards electrochemical nanoarchitectonics in energy-related applications. *Bull. Chem. Soc. Jpn.* **2017**, *90*, 627–648. [CrossRef]
15. Lin, C.; Hu, L.; Cheng, C.; Sun, K.; Guo, X.; Shao, Q.; Li, J.; Wang, N.; Guo, Z. Nano-$TiNb_2O_7$/carbon nanotubes composite anode for enhanced lithium-ion storage. *Electrochem. Acta* **2018**, *260*, 65–72. [CrossRef]
16. Murat, C.; Kakarla, R.R.; Fernando, A.M. Advanced electrochemical energy storage supercapacitors based on the flexible carbon fiber fabric-coated with uniform coral-like MnO_2 structured electrodes. *Chem. Eng. J.* **2017**, *309*, 151–158.
17. Gu, W.; Yushin, G. Review of nanostructured carbon materials for electrochemical capacitor applications: Advantages and limitations of activated carbon, carbide-derived carbon, zeolite-templated carbon, carbon aerogels, carbon nanotubes, onion-like carbon, and graphene. *WIREs Energy Environ.* **2014**, *3*, 424–473. [CrossRef]
18. Zeiger, M.; Jackel, N.; Mochalin, V.N.; Presser, V. Review: Carbon onions for electrochemical energy storage. *J. Mater. Chem. A* **2016**, *4*, 3172–3196. [CrossRef]
19. Wang, X.; Chen, S.; Li, D.; Sun, S.; Peng, Z.; Komarneni, S.; Yang, D. Direct interfacial growth of MnO_2 nanostructure on hierarchically porous carbon for high-performance asymmetric supercapacitors. *ACS Sustain. Chem. Eng.* **2018**, *6*, 633–641. [CrossRef]
20. Conway, B.E. Transition from "supercapacitor" to "battery" behavior in electrochemical energy storage. *J. Electrochem. Soc.* **1991**, *138*, 1539–1548. [CrossRef]
21. Cottineau, T.; Toupin, M.; Delahaye, T.; Brousse, T.; Belanger, D. Nanostructured transition metal oxides for aqueous hybrid electrochemical supercapacitors. *Appl. Phys. A Mater. Sci. Process.* **2006**, *82*, 599–606. [CrossRef]
22. Qu, Q.; Zhang, P.; Wang, B.; Chen, Y.; Tian, S.; Wu, Y.; Holze, R. Electrochemical performance of MnO_2 nanorods in neutral aqueous electrolytes as a cathode for asymmetric supercapacitors. *J. Phys. Chem. C* **2009**, *113*, 14020–14027. [CrossRef]
23. Yan, J.; Fan, Z.; Sun, W.; Ning, G.; Wei, T.; Zhang, Q.; Zhang, R.; Zhi, L.; Wei, F. Advanced asymmetric supercapacitors based on $Ni(OH)_2$/graphene and porous graphene electrodes with high energy density. *Adv. Funct. Mater.* **2012**, *22*, 2632–2641. [CrossRef]
24. Yang, P.; Ding, Y.; Lin, Z.; Chen, Z.; Li, Y.; Qiang, P.; Ebrahimi, M.; Mai, W.; Wong, C.P.; Wang, Z.L. Low-cost high-performance solid-state asymmetric supercapacitors based on MnO_2 nanowires and Fe_2O_3 nanotubes. *Nano Lett.* **2014**, *14*, 731–736. [CrossRef] [PubMed]
25. Jiang, H.; Lee, P.S.; Li, C.Z. 3D carbon based nanostructures for advanced supercapacitors. *Energy Environ. Sci.* **2013**, *6*, 41–52. [CrossRef]
26. Li, H.H.; Zhang, X.D.; Ding, R.; Qi, L.; Wang, H.Y. Facile synthesis of mesoporous MnO_2 microspheres for high performance AC//MnO_2 aqueous hybrid supercapacitor. *Electrochim. Acta* **2013**, *108*, 497–505. [CrossRef]
27. Shimamoto, K.; Tadanaga, K.; Tatsumisago, M. All-solid-state electrochemical capacitors using MnO_2/carbon nanotube composite electrode. *Electrochim. Acta* **2013**, *109*, 651–655. [CrossRef]
28. Wang, H.J.; Peng, C.; Zheng, J.D.; Peng, F.; Yu, H. Design, Synthesis and the electrochemical performance of MnO_2/C@CNT as supercapacitor material. *Mater. Res. Bull.* **2013**, *48*, 3389–3393. [CrossRef]
29. Devaraj, S.; Munichandraiah, N. Effect of crystallographic structure of MnO_2 on its electrochemical capacitance properties. *J. Phys. Chem. C* **2008**, *112*, 4406–4417. [CrossRef]
30. Xiao, X.; Li, T.; Yang, P.; Gao, Y.; Jin, H.; Ni, W.; Zhan, W.; Zhang, X.; Cao, Y.; Zhong, J.; et al. Fiber-based all-solid-state flexible supercapacitors for self-powered systems. *ACS Nano* **2012**, *6*, 9200–9206. [CrossRef] [PubMed]
31. Jiang, H.; Dai, Y.; Hu, Y.; Chen, W.; Li, C. Nanostructured ternary nanocomposite of rGO/CNTs/MnO_2 for high-rate supercapacitors. *ACS Sustain. Chem. Eng.* **2014**, *2*, 70–74. [CrossRef]
32. Liu, J.; Zhang, Y.; Li, Y.; Li, J.; Chen, Z.; Feng, H.; Li, J.; Jiang, J.; Qian, D. In situ chemical synthesis of sandwich-structured MnO_2/graphene nanoflowers and their supercapacitive behavior. *Electrochim. Acta* **2015**, *173*, 148–155. [CrossRef]
33. Yu, N.; Yin, H.; Zhang, W.; Liu, Y.; Tang, Z.; Zhu, M.Q. High-performance fiber-shaped all-solid-state asymmetric supercapacitors based on ultrathin MnO_2 nanosheet/carbon fiber cathodes for wearable electronics. *Adv. Energy Mater.* **2016**, *6*, 1501458. [CrossRef]

34. Teng, F.; Santhanagopalan, S.; Wang, Y.; Meng, D.D. In-situ hydrothermal synthesis of three-dimensional MnO_2-CNT nanocomposites and their electrochemical properties. *J. Alloys Compd.* **2010**, *499*, 259–264. [CrossRef]
35. Li, G.R.; Feng, Z.P.; Ou, Y.N.; Wu, D.; Fu, R.; Tong, Y.X. Mesoporous MnO_2/carbon aerogel composites as promising electrode materials for high-performance supercapacitors. *Langmuir* **2010**, *26*, 2209–2213. [CrossRef] [PubMed]
36. Li, L.; Hu, Z.A.; An, N.; Yang, Y.Y.; Li, Z.M.; Wu, H.Y. Facile synthesis of MnO_2/CNTs composite for supercapacitor electrodes with long cycle stability. *J. Phys. Chem. C* **2014**, *118*, 22865–22872. [CrossRef]
37. Zhu, K.; Wang, Y.; Tang, J.; Qiu, H.L.; Meng, X.; Gao, Z.M.; Chen, G.; Wei, Y.J.; Gao, Y. In situ growth of MnO_2 nanosheets on activated carbon fibers: A low-cost electrode for high performance supercapacitors. *RSC Adv.* **2016**, *6*, 14819–14825. [CrossRef]
38. Xu, W.B.; Mu, B.; Zhang, W.B.; Wang, A.Q. Facile hydrothermal synthesis of tubular kapok fiber/MnO_2 composites and application in supercapacitors. *RSC Adv.* **2015**, *5*, 64065–64075. [CrossRef]
39. Jin, Y.; Chen, H.Y.; Chen, M.H.; Liu, N.; Li, Q.W. Graphene patched CNT/MnO_2 nanocomposite papers for the electrode of high-performance flexible asymmetric supercapacitors. *ACS Appl. Mater. Interfaces* **2013**, *5*, 3408–3416. [CrossRef] [PubMed]
40. Li, M.; Tang, Z.; Mei, L.; Xue, J. Flexible solid-state supercapacitor based on graphene-based hybrid films. *Adv. Funct. Mater.* **2014**, *24*, 7495–7502. [CrossRef]
41. Wu, Z.S.; Wang, D.W.; Ren, W.; Zhao, J.; Zhou, G.; Li, F.; Cheng, H.M. Anchoring hydrous RuO_2 on graphene sheets for high-performance electrochemical capacitors. *Adv. Funct. Mater.* **2010**, *20*, 3595–3602. [CrossRef]
42. Zeng, W.; Zhang, G.; Hou, S.; Wang, T.; Duan, H. Facile synthesis of graphene@NiO/MoO_3 composite nanosheet arrays for high-performance supercapacitors. *Electrochim. Acta* **2015**, *151*, 510–516. [CrossRef]
43. Wu, Z.S.; Ren, W.; Wen, L.; Gao, L.; Zhao, J.; Chen, Z.; Zhou, G.; Li, F.; Cheng, H.M. Graphene anchored with Co_3O_4 nanoparticles as anode of lithium ion batteries with enhanced reversible capacity and cyclic performance. *ACS Nano* **2010**, *4*, 3187–3194. [CrossRef] [PubMed]
44. Li, H.; Jiang, L.; Cheng, Q.; He, Y.; Pavlinek, V.; Saha, P.; Li, C. MnO_2 nanoflakes/hierarchical porous carbon nanocomposites for high-performance supercapacitor electrodes. *Electrochim. Acta* **2015**, *164*, 252–259. [CrossRef]
45. Chen, Y.; Xu, J.; Yang, Y.; Zhao, Y.; Yang, W.; He, X.; Li, S.; Jia, C. Enhanced electrochemical performance of laser scribed graphene films decorated with manganese dioxide nanoparticles. *J. Mater. Sci. Mater. Electron.* **2016**, *27*, 2564–2573. [CrossRef]
46. Ghasemi, S.; Hosseinzadeh, R.; Jafari, M. MnO_2 nanoparticles decorated on electrophoretically deposited graphene nanosheets for high performance supercapacitor. *Int. J. Hydrogen Energy* **2015**, *40*, 1037–1046. [CrossRef]
47. Zheng, Y.; Pann, W.; Zhengn, D.; Sun, C. Fabrication of functionalized graphene- based MnO_2 nanoflower through electrodeposition for high-performance supercapacitor electrodes. *J. Electrochem. Soc.* **2016**, *163*, D230–D238. [CrossRef]
48. Kang, Y.; Cai, F.; Chen, H.; Chen, M.; Zhang, R.; Li, Q. Porous reduced graphene oxide wrapped carbon nanotube-manganese dioxide nanocables with enhanced electrochemical capacitive performance. *RSC Adv.* **2015**, *5*, 6136–6141. [CrossRef]
49. Zhang, Z.; Xiao, F.; Qian, L.; Xiao, J.; Wang, S.; Liu, Y. Facile synthesis of 3D MnO_2-graphene and carbon nanotube-graphene composite networks for high-performance flexible, all-solid-state asymmetric supercapacitors. *Adv. Energy Mater.* **2014**, *4*, 1400064. [CrossRef]
50. Lin, M.; Chen, B.; Wu, X.; Qian, J.; Fei, L.; Lu, W.; Chan, L.W.H.; Yuan, J. Controllable in situ synthesis of epsilon manganese dioxide hollow structure/RGO nanocomposites for high-performance supercapacitors. *Nanoscale* **2016**, *8*, 1854–1860. [CrossRef] [PubMed]
51. Brousse, T.; Taberna, P.L.; Crosnier, O.; Dugas, R.; Guillemet, P.; Scudeller, Y.; Zhou, Y.; Favier, F.; Bélanger, D.; Simon, P. Long-term cycling behavior of asymmetric activated carbon/MnO_2 aqueous electrochemical supercapacitor. *J. Power Sources* **2007**, *173*, 633–641. [CrossRef]
52. Liu, Y.; Miao, X.; Fang, J.; Zhang, X.; Chen, S.; Li, W.; Feng, W.; Chen, Y.; Wang, W.; Zhang, Y. Layered-MnO_2 nanosheet grown on nitrogen-doped graphene template as a composite cathode for flexible solid-state asymmetric supercapacitor. *ACS Appl. Mater. Interfaces* **2016**, *8*, 5251–5260. [CrossRef] [PubMed]

53. Lu, X.; Yu, M.; Zhai, T.; Wang, G.; Xie, S.; Liu, T.; Liang, C.; Tong, Y.; Li, Y. High energy density asymmetric quasi-solid-state supercapacitor based on porous vanadium nitride nanowire anode. *Nano Lett.* **2013**, *13*, 2628–2633. [CrossRef] [PubMed]
54. Lu, X.; Yu, M.; Wang, G.; Zhai, T.; Xie, S.; Ling, Y.; Tong, Y.; Li, Y. H-TiO_2@ MnO_2//H-TiO_2@C core-shell nanowires for high performance and flexible asymmetric supercapacitors. *Adv. Mater.* **2013**, *25*, 267–272. [CrossRef] [PubMed]
55. Chen, P.; Chen, H.; Qiu, J.; Zhou, C. Inkjet printing of single-walled carbon nanotube/RuO_2 nanowire supercapacitors on cloth fabrics and flexible substrates. *Nano Res.* **2010**, *3*, 594–603. [CrossRef]
56. Hummers, W.S.; Offeman, R.E. Preparation of graphitic oxide. *J. Am. Chem. Soc.* **1958**, *80*, 1339. [CrossRef]
57. Kovtyukhova, N.I.; Ollivier, P.J.; Martin, B.R.; Mallouk, T.E.; Chizhik, S.A.; Buzaneva, E.V.; Gorchinskiy, A.D. Layer-by-layer assembly of ultrathin composite films from micron-sized graphite oxide sheets and polycations. *Chem. Mater.* **1999**, *11*, 771–778. [CrossRef]
58. Wen, Y.; Ding, H.; Shan, Y. Preparation and visible light photocatalytic activity of Ag/TiO_2/graphene nanocomposite. *Nanoscale* **2011**, *3*, 4411–4417. [CrossRef] [PubMed]
59. Zhang, Z.; Xiao, F.; Guo, Y.; Wang, S.; Liu, Y. One-pot self-assembled three-dimensional TiO_2-graphene hydrogel with improved adsorption capacities and photocatalytic and electrochemical activities. *ACS Appl. Mater. Interfaces* **2013**, *5*, 2227–2233. [CrossRef] [PubMed]
60. Chee, W.K.; Lim, H.N.; Zainal, Z.; Huang, N.M.; Harrison, I.; Andou, Y. Flexible graphene-based supercapacitors: A review. *J. Phys. Chem. C* **2016**, *120*, 4153–4172. [CrossRef]
61. Liu, L.; Su, L.; Lang, J.; Hu, B.; Xu, S.; Yan, X. Controllable synthesis of Mn_3O_4 nanodots@nitrogen-doped graphene and its application for high energy density supercapacitors. *J. Mater. Chem. A* **2017**, *5*, 5523–5531. [CrossRef]
62. Xu, H.; Qu, Z.; Zong, C.; Huang, W.; Quan, F.; Yan, N. MnO_x/graphene for the catalytic oxidation and adsorption of elemental mercury. *Environ. Sci. Technol.* **2015**, *49*, 6823–6830. [CrossRef] [PubMed]
63. Yang, Y.; Zeng, B.; Liu, J.; Long, Y.; Li, N.; Wen, Z.; Jiang, Y. Graphene/MnO_2 composite prepared by a simple method for high performance supercapacitor. *Mater. Res. Innov.* **2016**, *20*, 92–98. [CrossRef]
64. Li, Z.; Mi, Y.; Liu, X.; Liu, S.; Yang, S.; Wang, J. Flexible graphene/MnO_2 composite papers for supercapacitor electrodes. *J. Mater. Chem.* **2011**, *21*, 14706–14711. [CrossRef]
65. Unnikrishnan, B.; Wu, C.W.; Chen, I.W.P.; Chang, H.T.; Lin, C.H.; Huang, C.C. Carbon dot-mediated synthesis of manganese oxide decorated graphene nanosheets for supercapacitor application. *ACS Sustain. Chem. Eng.* **2016**, *4*, 3008–3016. [CrossRef]
66. Du, D.; Li, P.; Ouyang, J. Nitrogen-doped reduced graphene oxide prepared by simultaneous thermal reduction and nitrogen doping of graphene oxide in air and its application as an electrocatalyst. *ACS Appl. Mater. Interfaces* **2015**, *7*, 26952–26958. [CrossRef] [PubMed]
67. Wang, B.; Qin, Y.; Tan, W.; Tao, Y.; Kong, Y. Smartly designed 3D N-doped mesoporous graphene for high-performance supercapacitor electrodes. *Electrochim. Acta* **2017**, *241*, 1–9. [CrossRef]
68. Chen, S.; Zhu, J.; Wu, X.; Han, Q.; Wang, X. Graphene oxide MnO_2 nanocomposite for supercapacitors. *ACS Nano* **2010**, *4*, 2822–2830. [CrossRef] [PubMed]
69. Qu, J.; Shi, L.; He, C.; Gao, F.; Li, B.; Hu, Q.Z.H.; Shao, G.; Wang, X.; Qiu, J. Highly efficient synthesis of graphene/MnO_2 hybrids and their application for ultrafast oxidative decomposition of methylene blue. *Carbon* **2014**, *66*, 485–492. [CrossRef]
70. Arunabha, G.; Hee, L.Y. Carbon-based electrochemical capacitors. *ChemSusChem* **2012**, *5*, 480–499.
71. Sun, L.; Wang, L.; Tian, C.; Tan, T.; Xie, Y.; Shi, K.; Li, M.; Fu, H. Nitrogen-doped graphene with high nitrogen level via a one-step hydrothermal reaction of graphene oxide with urea for superior capacitive energy storage. *RSC Adv.* **2012**, *2*, 4498–4506. [CrossRef]
72. Zhang, W.; Xu, C.; Ma, C.; Li, G.; Wang, Y.; Zhang, K.; Li, F.; Liu, C.; Cheng, H.M.; Du, Y.; et al. Nitrogen-superdoped 3D graphene networks for high-performance supercapacitors. *Adv. Mater.* **2017**, *29*, 1701677. [CrossRef] [PubMed]
73. Zhang, X.; He, M.; He, P.; Li, C.; Liu, H.; Zhang, X.; Ma, Y. Ultrafine nano-network structured bacterial cellulose as reductant and bridging ligands to fabricate ultrathin K-birnessite type MnO_2 nanosheets for supercapacitors. *Appl. Surf. Sci.* **2018**, *433*, 419–427. [CrossRef]

74. Ren, Y.; Xu, Q.; Zhang, J.; Yang, H.; Wang, B.; Yang, D.; Hu, J.; Liu, Z. Functionalization of biomass carbonaceous aerogels: Selective preparation of MnO_2@CA composites for supercapacitors. *ACS Appl. Mater. Interfaces* **2014**, *6*, 9689–9697. [CrossRef] [PubMed]
75. Zolfaghari, A.; Naderi, H.R.; Mortaheb, H.R. Carbon black/manganese dioxide composites synthesized by sonochemistry method for electrochemical supercapacitors. *J. Electroanal. Chem.* **2013**, *697*, 60–67. [CrossRef]
76. Naderia, H.R.; Norouzia, P.; Ganjali, M.R. Electrochemical study of a novel high performance supercapacitor based on MnO_2/nitrogen-doped graphene nanocomposite. *Appl. Surf. Sci.* **2016**, *366*, 552–560. [CrossRef]
77. Li, Y.; Zhao, N.; Shi, C.; Liu, E.; He, C. Improve the supercapacity performance of MnO_2-decorated graphene by controlling the oxidization extent of graphene. *J. Phys. Chem. C* **2012**, *116*, 25226–25232. [CrossRef]
78. Huang, B.; Huang, C.; Qian, Y. Facile hydrothermal synthesis of manganese dioxide/nitrogen-doped graphene composites as electrode material for supercapacitors. *J. Electrochem. Sci.* **2017**, *12*, 11171–11180. [CrossRef]
79. Xiao, W.; Zhou, W.J.; Yu, H.; Pu, Y.; Zhang, Y.H.; Hu, C.B. Template synthesis of hierarchical mesoporous δ-MnO_2 hollow microspheres as electrode material for high-performance symmetric supercapacitor. *Electrochim. Acta* **2018**, *264*, 1–11. [CrossRef]

Communication

Single-Step Direct Hydrothermal Growth of $NiMoO_4$ Nanostructured Thin Film on Stainless Steel for Supercapacitor Electrodes

V. Kannan [1], Hyun-Jung Kim [2], Hyun-Chang Park [2] and Hyun-Seok Kim [2,*]

[1] Millimeter-Wave Innovation Technology Research Center (MINT), Dongguk University-Seoul, Seoul 04620, Korea; kannan@dongguk.edu

[2] Division of Electronics and Electrical Engineering, Dongguk University-Seoul, Seoul 04620, Korea; best7hj@daum.net (H.-J.K.); hcpark@dongguk.edu (H.-C.P.)

* Correspondence: hyunseokk@dongguk.edu; Tel.: +82-2-2260-3996; Fax: +82-2-2277-8735

Received: 24 June 2018; Accepted: 21 July 2018; Published: 24 July 2018

Abstract: We report a facile and direct growth of $NiMoO_4$ nanostructures on a nonreactive stainless steel substrate using a single-step hydrothermal method and investigated hydrothermal growth duration effects on morphology and electrochemical characteristics. The highest specific capacitances of 341, 619, and 281 F/g were observed for $NiMoO_4$ with 9, 18, and 27 h growth, respectively, at 1 A/g. Thus, grown samples preserved almost 59% of maximum specific capacitance at a high current density of 10 A/g. All samples exhibited a respectable cycling stability over 3000 charge-discharge operations. $NiMoO_4$ grown for 18 h exhibited 7200 W/kg peak power density at 14 Wh/kg energy density. Thus, the proposed single-step hydrothermal growth is a promising route to obtain $NiMoO_4$ nanostructures and other metal oxide electrodes for supercapacitor applications.

Keywords: $NiMoO_4$; nanostructures; hydrothermal; supercapacitor; stainless steel

1. Introduction

Supercapacitors have been widely studied as candidates for energy storage systems due to their large power density, rapid charge-discharge, excellent rate capabilities, and good endurance [1–3]. They have been used in tandem with rechargeable storage devices and fuel cells, and they have been broadly applied for electric vehicles, electrical grid buffers, space applications, and memory system power back-up. In the recent past, there has been significant surge of research in transition metal oxides-based supercapacitors due to their multiple oxidation states and reversible redox reaction capabilities [4–6]. However, progress in supercapacitor electrode materials has been constrained by high cost (e.g., RuO_2) or environmental impact (e.g., metal sulfides) [7–9].

Metal oxides with two transition metals have also been studied to take advantage of their variable oxidation states, excellent electrical conduction, and enhanced pseudo-capacitance characteristics, such as $CoMoO_4$ [10,11], $NiCo_2O_4$ [6], $ZnWO_4$ [12], and $NiMoO_4$ [13]. $NiMoO_4$ is a promising binary metal oxide due to its potential for high specific capacitance arising from the excellent electrochemical nature of the Ni ion, although practical electrode use is deterred by its poor electrical conductivity. Intricate characteristics of the molybdate also complicate $NiMoO_4$ nanostructure growth [14,15]. Hence, it is important to identify easier preparation routes for $NiMoO_4$ nanostructure thin films with distinct morphologies and superior supercapacitor characteristics.

There is a promising route to grow the nanostructures directly on conducting substrates [16], countering inherent low metal oxide electrical conductivity and, hence, considerably enhancing electrochemical performance. There are many reports of $NiMoO_4$ supercapacitor performance on Ni foam substrates [14–18]. However, the $NiMoO_4$ on Ni foam (usually coated with a binder as a slurry) is often dominated by the Ni foam contribution to overall electrochemical performance, making it difficult to identify the exact supercapacitance value of the desired electrode material. Only $NiMoO_4$ nanostructure synthesis on nonreactive substrates has been reported previously, such as stainless steel (SS) and carbon [16,19], without exploring its electrochemical characteristics. However, there are no reports available detailing the electrochemical performance of $NiMoO_4$ thin film nanostructures on nonreactive substrate such as stainless steel. Here we intend to exclusively evaluate the electrochemical performance of $NiMoO_4$ thin film nanostructures, without the hindrance of substrate contribution.

This study reports on the binder-free growth of $NiMoO_4$ nanostructures on SS substrate using a facile, single-step hydrothermal technique. The optimal growth time was identified by studying $NiMoO_4$ thin films grown for 9, 18, and 27 h, keeping other deposition conditions constant. Nanostructured $NiMoO_4$ grown on SS substrate exhibited high specific capacitance, good cycling stability, and enhanced rate capability. The proposed technique offers a promising, environmentally friendly, and relatively low cost direct route to obtain high supercapacitance $NiMoO_4$ nanostructures.

2. Materials and Methods

2.1. *$NiMoO_4$ Nanostructure Growth Process*

Molybdenum chloride (1.5 mmol) was mixed with methanol (50 mL) and stirred for 10 min. Nickel chloride solution (1.5 mmol) was added and stirred for a further 10 min, then EDTA (1.5 mmol) was added and stirred for 1 h. After forming a clear solution, 1 mL H_2O_2 and 1 mL HNO_3 were added and stirred for 10 min. The solution was transferred to a Teflon container that already contained a pre-cleaned SS substrate with 1 cm × 2 cm exposed area. The complete setup was placed in a stainless steel autoclave at 180 °C. In order to identify optimal growth time, experiments were conducted with wide ranging growth durations from 9 to 36 h. Three samples were identified with 9, 18, and 27 h growth duration and labeled as NMO-9, NMO-18, and NMO-27, respectively, due to their superior electrochemical performance over the other samples. The grown films are then harvested, washed with deionized water, and dried with N_2 gas.

2.2. *Materials Characterization*

The obtained thin films were characterized using a field emission scanning electron microscope (FE-SEM, Hitachi-S-4800, Huntington Beach, CA, USA), transmission electron microscope (TEM, JEM-2100F, JEOL, Akishima, Tokyo, Japan), and high angle annular dark field imaging (HAADF) scanning transmission electron microscope (STEM, JEM-2100F, JEOL, Akishima, Tokyo, Japan). Electrochemical measurements were performed in a 2 M KOH aqueous solution using a standard three-electrode electrochemical cell in Versa-stat-3. $NiMoO_4$ served as the working electrode, with a saturated calomel electrode (SCE) and graphite rod as reference and counter electrodes, respectively.

3. Results

FE-SEM morphology of the grown films for the durations of 9 h, 18 h, and 27 h are shown in Figure 1. $NiMoO_4$ nanostructures are evident for all the films grown for different durations, exhibiting a stacked structure with $NiMoO_4$ nanograins. TEM analyses were carried out only on sample NMO-18 due to its superior electrochemical performance in comparison with NMO-9 and NMO-27. Figure 2a,b show typical TEM images for the 18 h sample, revealing crystalline $NiMoO_4$ grain-like structure. From the observed twist dislocations in Figure 2b, it can be inferred that the crystals are composed of narrow platelets. Figure 2c shows a typical high resolution transmission electron microscope (HRTEM) image from a single nanograin. The interplanar distance (0.39 nm) corresponds to monoclinic $NiMoO_4$ $(02\bar{1})$ crystal planes (JCPDS: 45-0142). Figure 2d shows the selected area diffraction (SAED) pattern, which confirms the polycrystalline nature. Figure 2e–h shows the HAADF-STEM electron, O, Ni, and Mo mappings, respectively, using the same scale is the TEM images. Ni, Mo, and O are relatively uniformly distributed, which is confirmed by the elemental line mapping, as shown in Figure 2i.

Figure 1. Typical field emission scanning electron microscope (FE-SEM) images for (**a**) NMO-9; (**b**) NMO-18; and (**c**) NMO-27 samples.

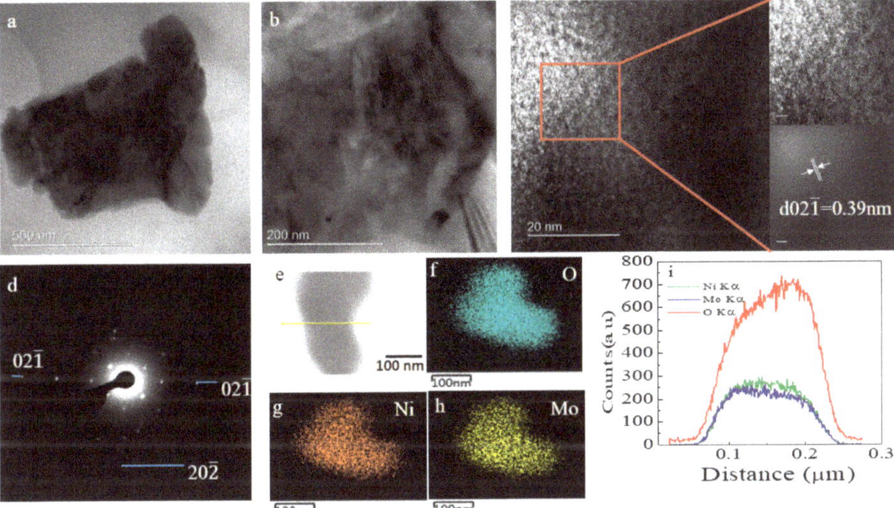

Figure 2. The transmission electron microscope (TEM) image for a single $NiMoO_4$ nanograin grown for 18 h at (**a**) low and (**b**) high magnification; (**c**) The high resolution transmission electron microscope (HRTEM) image with d spacing noted; (**d**) The selected area diffraction (SAED) pattern of the same sample showing respective planes. High angle annular dark field imaging scanning transmission electron microscope (HAADF-STEM) elemental mapping for a selected area showing (**e**) the electron with line mapping; (**f**) oxygen; (**g**) nickel; (**h**) molybdenum; and (**i**) line mapping, showing the counts of Mo, Ni, and O.

The electrochemical characteristics of the electrode are mainly dependent on their dimensions and morphologies [20,21]. Electrochemical measurements of $NiMoO_4$ nanostructure were performed in 2 M KOH electrolyte. Figure 3a shows the cyclic voltammograms (CV) for hydrothermally grown $NiMoO_4$ at different durations (100 mV/s scan rate), over the potential range of 0–0.43 V (versus SCE). Each CV plot exhibits distinct redox peaks, indicating that the observed capacitance properties can be described by Faradic reactions [22,23]. The NMO-18 sample had the largest area under the CV curve compared with NMO-9 and NMO-27.

Figure 3b–d show NMO-9, NMO-18, and NMO-27 CVs, respectively, for various scan rates. The CV curve shapes are almost unchanged for the different scan rates, indicating near ideal capacitive characteristics. Voltage corresponding to the oxidation peak moved in the positive direction as the scan rate increased, whereas the reduction peak moved in the negative direction, which can be attributed to electrode's internal resistance [24].

The surface area accessible for electrochemical reactions can be estimated from the electrochemically active surface area (ECSA). Initially, the non-Faradic capacitive current was obtained from the linear region of the CV curves,

$$i_{DL} = C_{DL} \times \nu \tag{1}$$

where i_{DL} is the capacitive current, C_{DL} is the specific capacitance in the non-Faradic region, and ν is the scan rate. Electrochemical capacitance was estimated for each sample at the various scan rates, and ECSA was calculated from

$$ECSA = C_{DL}/C_s \tag{2}$$

where C_s is the specific capacitance of the standard electrode in an alkaline electrolyte [25]. We chose the non-Faradic region as 0.21–0.23 V, as shown in Figure 4a. Figure 4b shows the i_{DL} at 0.22 V for the scan rate, with a corresponding ECSA of 649, 874, and 342 cm^{-2} for NMO-9, NMO-18, and NMO-27, respectively. NMO-18 exhibited the highest ECSA, consistent with its observed maximum supercapacitance. Thus, hydrothermal growth duration is critical to obtain the largest surface area and, hence, the highest supercapacitance.

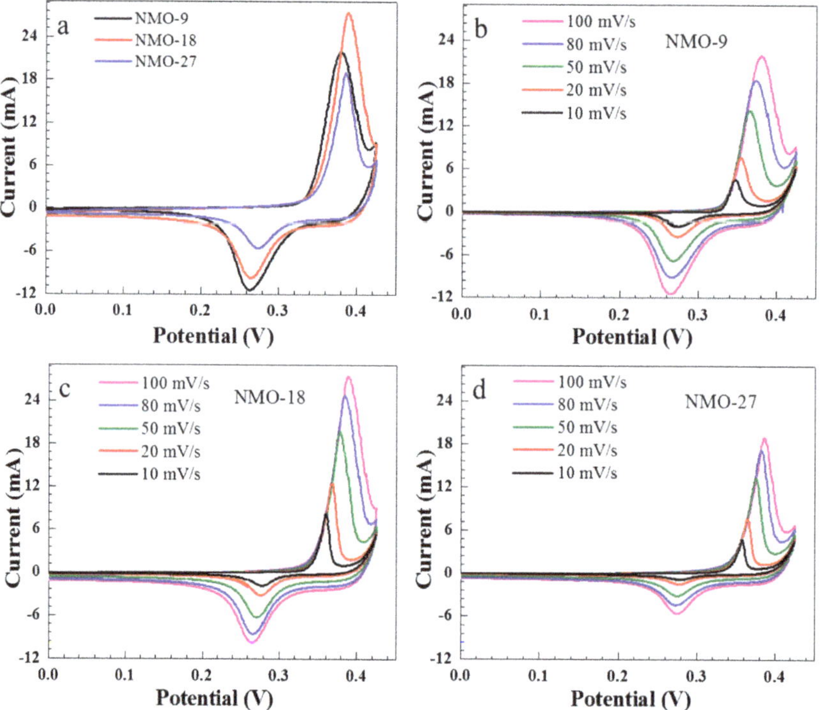

Figure 3. Cyclic voltammograms (CVs) for (**a**) NMO-9, NMO-18, and NMO-27 samples at 100 mV/s scan rate and various scan rates for (**b**) NMO-9; (**c**) NMO-18; and (**d**) NMO-27.

Figure 5a shows the galvanostatic charge-discharge profiles for the $NiMoO_4$ nanostructure electrodes at a 1 A/g current density, with NMO-18 exhibiting the longest discharge time. Figure 5b–d show NMO-9, NMO-18, and NMO-27 charge-discharge profiles, respectively, at different current densities.

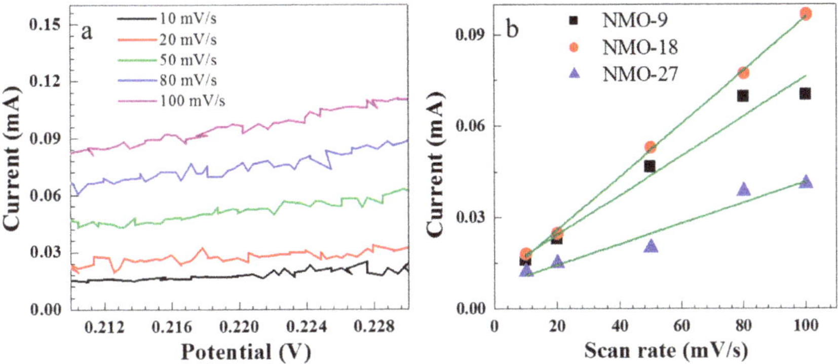

Figure 4. (**a**) Cyclic voltammograms (CV) for NMO-18 at various scan rates in the non-Faradic voltage region (0.21–0.23 V); (**b**) The measured current at 0.22 V as a function of the scan rate for NMO-9, NMO-18, and NMO-27.

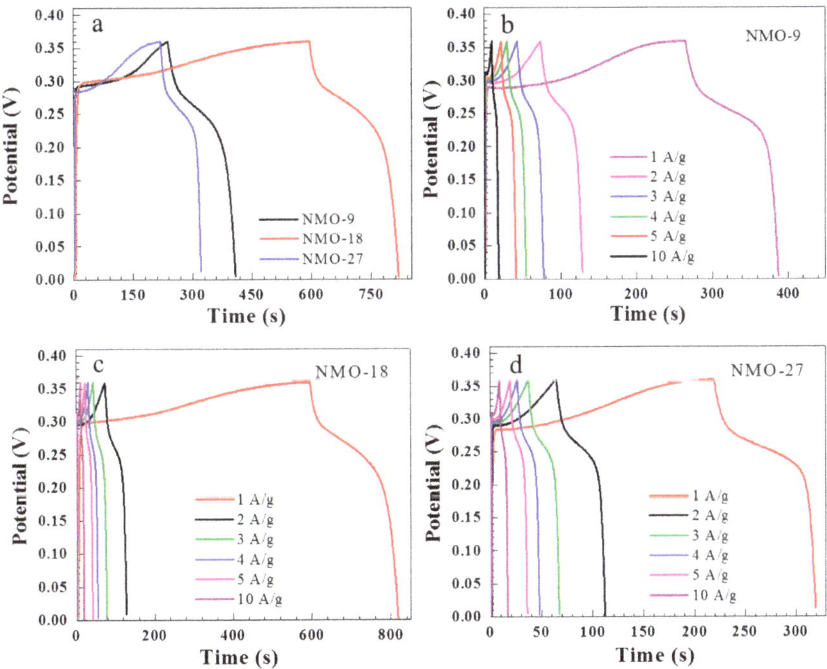

Figure 5. Galvanostatic charge-discharge profile for (**a**) NMO-9, NMO-18, and NMO-27 at 1 A/g current density; (**b**) NMO-9; (**c**) NMO-18; and (**d**) NMO-27 at various current densities.

Each charge-discharge curve shows pseudo-capacitor characteristics, consistent with the CV data. There is a sharp voltage drop as the supercapacitor changes state from charge to discharge, which can be attributed to an IR drop. Each electrode exhibits nonlinear discharge followed by a plateau, which verifies faradaic reactions occurring on the electrode surface due to the redox reaction at the material/solution interface. Figure 6a represents the specific capacitance response for current densities for all samples. The performance of the specific capacitance decreases with increasing current density. The fall in the supercapacitance value at a high current density could be attributed to the lack of active material at the electrode/electrolyte interface during oxidation/reduction reactions and the slow diffusion of bulky OH ions [26,27].

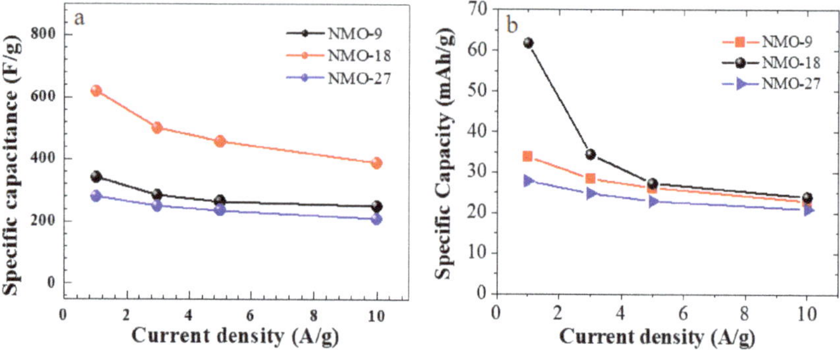

Figure 6. (**a**) Specific capacitance as a function of current density for all the samples of NMO-9, NMO-18, and NMO-27; (**b**) The response of specific capacity to current density for NMO-9, NMO-18, and NMO-27.

Specific capacitance or C_s was calculated from galvanostatic charge-discharge plots (Figure 5) from

$$C_s = \frac{I \cdot \Delta t}{m \cdot \Delta V}, \quad (3)$$

where I is the applied current density, ΔV is the potential window, and Δt is the discharge time. The calculated specific capacitances at 1 A/g are 341, 619, and 219 F/g for NMO-9, NMO-18, and NMO-27, respectively. The observed specific capacitances are directly related to the available surface area for the electrochemical reaction in the $NiMoO_4$ nanostructure electrode. The superior specific capacitance observed in NMO-18, in comparison with NMO-9 and NMO-27, can be attributed to the higher electrochemical surface area observed in NMO-18 from the ECSA estimation described above. Thus, $NiMoO_4$ nanostructures showed comparable capacitance and rate capability with previously reported materials, such as NiO [26,28], Co_3O_4 [20], and MnO_2 [29]. Table 1 shows the super capacitance values of this work along with other $NiMoO_4$ reports on various nanostructures and different substrates. The specific capacity was also calculated due to the fact that CV-plots exhibit sharp redox and oxidation peaks arising from a Faradic battery-type mechanism [30,31]. Therefore, we also present specific capacity, C (mAh/g), which can be calculated from charge-discharge plots using,

$$C = \frac{I \cdot \Delta t}{3600 \cdot m} \quad (4)$$

The specific capacity for all the samples at various current densities are shown in Figure 6b. The specific capacity for NMO-9, NMO-18, and NMO-27 are 33.9, 61.8, and 27.9 mAh/g, respectively, at a current density of 1 A/g.

Table 1. Data of previous $NiMoO_4$ nanostructure-based reports compared with our work.

S. No.	$NiMoO_4$ Electrode Type	Substrate	Substrate Type	Super Capacitance (F/g)	Reference
1.	Nanorods	Ni-Foam	Reactive	1136	[16]
2.	Nanorods	Ni-Foam	Reactive	944	[17]
3.	Nanotubes	Ni-Foam	Reactive	864	[18]
4.	Nanoneedles	Carbon	Non-reactive	412	[19]
5.	Nanograins	Stainless Steel	Non-reactive	619	This work

We also study electrochemical stability, an important characteristic for practical applications. Each $NiMoO_4$ electrode was subjected to 3000 charge-discharge cycles. Figure 7a shows the electrode stability at 10 A/g current density. NMO-18 exhibited a steep capacitance loss during the initial several hundred cycles, and then stabilized to approximately 57% retention after 3000 cycles [32]. NMO-9 and NMO-27 showed 44% and 56.5% retention after 3000 cycles of charge-discharge operations, respectively. The lower retention level in NMO-9 may be attributed to a more rapid attrition of electrode material in relation to other two samples during charge-discharge cycles. The capacitance reduction can be explained by $NiMoO_4$ material physical expansion as ionic transfer occurred and the partial dissolution of $NiMoO_4$ material during charge-discharge operations [33].

Energy density (E) and power density (P) can be expressed as

$$E = \frac{1}{2} C_S \cdot (\Delta V)^2 \tag{5}$$

and

$$P = \frac{E}{\Delta t} \tag{6}$$

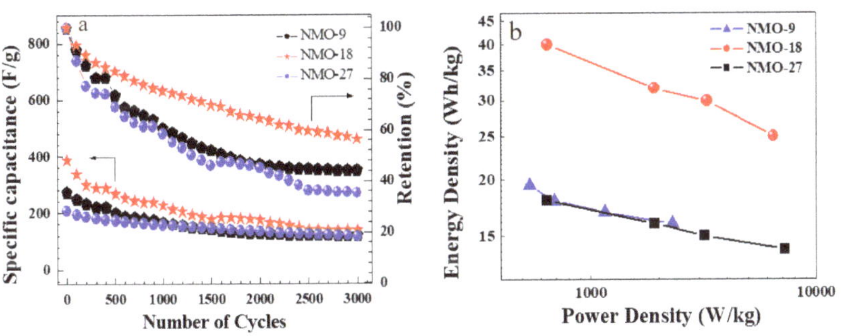

Figure 7. (**a**) Response of specific capacitance over 3000 cycles of charge-discharge operations for NMO-9, NMO-18, and NMO-27; (**b**) Ragone plots with energy density versus power density.

Figure 7b shows the Ragone plots comparing energy and power density for all the electrodes. The peak energy density was observed for NMO-18 as 40 Wh/kg, and the highest power density of 7200 W/kg was associated with NMO-27. In a largely diffusion-controlled redox reaction between NiO and OH^-, it is expected that energy density and charge-discharge rates will be inversely proportional [34–36]. It is notable that the $NiMoO_4$ nanostructures exhibited high power density and energy density, consistent with past research reports [37,38]. Table 2 shows the specific capacitance and energy density found in the current work in relation to the previously published metal oxide reports.

Table 2. Electrochemical performance of this work compared with previously reported various metal oxide supercapacitor electrode materials.

S. No.	Electrode Material	Substrate	Specific Capacitance (F/g)	Energy Density (Wh/kg)	Reference
1.	PANI/PTSa/cross linked $NiMoO_4$	Graphite	1300	60	[39]
2.	$NiMoO_4$ $CoMoO_4$ $MnMoO_4$	Carbon	412 240 120	35 15 10	[19]
3.	$CoMoO_4$	Carbon	79	21	[40]
4.	$NiMn_2O_4$/CNT	Ni	151	60.69	[41]
5.	RuO_2	-	658	-	[42]
6.	MnO_2	-	1380	-	[43]
7.	$NiMoO_4$	SS	619	40	This work

4. Conclusions

Direct $NiMoO_4$ nanostructure growth on SS substrates was achieved via a facile, single-step hydrothermal method, and the resultant electrode electrochemical characteristics were investigated with respect to growth times of 9, 18, and 27 h. The NMO-18 electrode exhibited the maximum supercapacitance value of 619, 500, 458, and 390 F/g at 1, 3, 5, and 10 A/g, respectively. All the electrodes showed good stability over 3000 charge-discharge cycles. The available active surface area for all the samples were estimated from ECSA analysis. The highest capacitance observed in NMO-18 could be attributed to the observed maximum ECSA value of the sample. Along with excellent capacitive characteristics, $NiMoO_4$ showed appreciable energy and power density values. We propose that the direct growth of nanostructured $NiMoO_4$ on SS substrate through a one-step hydrothermal method offers a promising electrode material for supercapacitor applications.

Author Contributions: V.K. did the experiment and prepared the manuscript. H.-J.K. helped with analysis and discussion. H.-C.P. helped in performing the discussion and revising the manuscript. H.-S.K. planned and supervised the project. All of the authors have discussed the results and approved the submitted version.

Funding: This work was supported by the Korea Institute of Energy Technology Evaluation and Planning (KETEP) and the Ministry of Trade, Industry & Energy (MOTIE) of the Republic of Korea (No. 20174030201520), and the Basic Science Research Program through the National Research Foundation of Korea (NRF) funded by the Ministry of Education (No. 2017R1D1A1A09000823).

Conflicts of Interest: There are no conflicts of interest to declare.

References

1. Miller, J.R.; Simon, P. Electrochemical Capacitors for Energy Management. *Science* **2008**, *321*, 651–652. [CrossRef] [PubMed]
2. Simon, P.; Gogotsi, Y. Materials for electrochemical capacitors. *Nat. Mater.* **2008**, *7*, 845–854. [CrossRef] [PubMed]
3. Liban, D.J.; Saleh, M.A.; Berkson, Z.J.; El-Kady, M.F.; Hwang, J.Y.; Mohamed, N.; Wixtrom, A.I.; Titarenko, E.; Shao, Y.; McCarthy, K.; et al. A molecular cross-linking approach for hybrid metal oxides. *Nat. Mater.* **2018**, *17*, 341–348.
4. Cheng, F.; Shen, J.; Peng, B.; Pan, Y.; Tao, Z.; Chen, J. Rapid room-temperature synthesis of nanocrystalline spinels as oxygen reduction and evolution electrocatalysts. *Nat. Chem.* **2011**, *3*, 79–84. [CrossRef] [PubMed]
5. Wu, H.B.; Pang, H.; Lou, X.W. Facile synthesis of mesoporous $Ni_{0.3}Co_{2.7}O_4$ hierarchical structures for high-performance supercapacitors. *Energy Environ. Sci.* **2013**, *6*, 3619–3626. [CrossRef]
6. Liang, Y.; Wang, H.; Zhou, J.; Li, Y.; Wang, J.; Regier, T.; Dai, H. Covalent hybrid of spinel manganese–cobalt oxide and graphene as advanced oxygen reduction electrocatalysts. *J. Am. Chem. Soc.* **2012**, *134*, 3517–3523. [CrossRef] [PubMed]
7. Zhang, J.T.; Ma, J.Z.; Zhang, L.L.; Guo, P.Z.; Jiang, J.W.; Zhao, X.S. Template synthesis of tubular ruthenium oxides for supercapacitor applications. *J. Phys. Chem. C* **2010**, *114*, 13608–13613. [CrossRef]
8. Wang, Q.H.; Jiao, L.F.; Du, H.M.; Peng, W.X.; Han, Y.; Song, D.W.; Si, Y.C.; Wang, Y.J.; Yuan, H.T. Novel flower-like CoS hierarchitectures: One-pot synthesis and electrochemical properties. *J. Mater. Chem.* **2011**, *21*, 327–329. [CrossRef]
9. Zhou, L.; He, Y.; Jia, C.; Pavlinek, V.; Saha, P.; Cheng, Q. Construction of hierarchical $CuO/Cu_2O@NiCo_2S_4$ nanowire arrays on copper foam for high performance supercapacitor electrodes. *Nanomaterials* **2017**, *7*, 273. [CrossRef] [PubMed]
10. Mai, L.Q.; Yang, F.; Zhao, Y.L.; Xu, X.; Xu, L.; Luo, Y.Z. Hierarchical $MnMoO_4/CoMoO_4$ heterostructured nanowires with enhanced supercapacitor performance. *Nat. Commun.* **2011**, *2*, 381. [CrossRef] [PubMed]
11. Liu, M.C.; Kang, L.; Kong, L.B.; Lu, C.; Ma, X.J.; Li, X.M.; Luo, Y.C. Facile synthesis of $NiMoO_4{\cdot}xH_2O$ nanorods as a positive electrode material for supercapacitors. *RSC Adv.* **2013**, *3*, 6472–6478. [CrossRef]
12. Guan, B.K.; Hu, L.L.; Zhang, G.H.; Guo, D.; Fu, T.; Li, J.D.; Duan, H.G.; Li, Q.H. Facile synthesis of $ZnWO_4$ nanowall arrays on Ni foam for high performance supercapacitors. *RSC Adv.* **2014**, *4*, 4212–4217. [CrossRef]
13. Guo, D.; Zhang, P.; Zhang, H.; Yu, X.; Zhu, J.; Li, Q.; Wang, T. $NiMoO_4$ nanowires supported on Ni foam as novel advanced electrodes for supercapacitors. *J. Mater. Chem. A* **2013**, *7*, 9024–9027. [CrossRef]

14. Yu, L.; Zhang, L.; Wu, H.B.; Zhang, C.; Lou, X.W. Controlled synthesis of hierarchical $Co_xMn_{3-x}O_4$ array micro-/nanostructures with tunable morphology and composition as integrated electrodes for lithium-ion batteries. *Energy Environ. Sci.* **2013**, *6*, 2664–2671. [CrossRef]
15. Shaijumon, M.M.; Perre, E.; Daffos, B.; Taberna, P.L.; Tarascon, J.M.; Simon, P. Nanoarchitectured 3D cathodes for Li-Ion microbatteries. *Adv. Mater.* **2010**, *22*, 4978–4981. [CrossRef] [PubMed]
16. Peng, S.; Li, L.; Wu, H.B.; Madhavi, S.; Lou, X.W. Controlled growth of $NiMoO_4$ nanosheet and nanorod arrays on various conductive substrates as advanced electrodes for asymmetric supercapacitors. *Adv. Energy Mater.* **2015**, *5*, 1401172. [CrossRef]
17. Cai, D.; Wang, D.; Liu, B.; Wang, Y.; Liu, Y.; Wang, L.; Li, H.; Huang, H.; Li, Q.; Wang, T. Comparison of the electrochemical performance of $NiMoO_4$ nanorods and hierarchical nanospheres for supercapacitor applications. *ACS Appl. Mater. Interface* **2013**, *5*, 12905–12910. [CrossRef] [PubMed]
18. Yin, Z.; Zhang, S.; Chen, Y.; Gao, P.; Zhu, C.; Yang, P.; Qi, L. Hierarchical nanosheet-based $NiMoO_4$ nanotubes: Synthesis and high supercapacitor performance. *J. Mater. Chem. A* **2015**, *3*, 739–745. [CrossRef]
19. Teeraphat, W.; Manickam, M.S.; Sudip, C.; Dan, L.; Shafiullah, G.M.; Robert, D.A.; Rajeev, A. Effect of transition metal cations on stability enhancement for molybdate-based hybrid supercapacitor. *ACS Appl. Mater. Interfaces* **2017**, *9*, 17977–17991.
20. Meher, S.K.; Rao, G.R. Ultralayered Co_3O_4 for high-performance supercapacitor applications. *J. Phys. Chem. C* **2011**, *115*, 15646–15654. [CrossRef]
21. Chen, J.S.; Ng, M.F.; Wu, H.B.; Zhang, L.; Lou, X.W. Synthesis of phase-pure SnO_2 nanosheets with different organized structures and their lithium storage properties. *CrystEngComm* **2012**, *14*, 5133–5136. [CrossRef]
22. Zhang, G.Q.; Lou, X.W. General solution growth of mesoporous $NiCo_2O_4$ nanosheets on various conductive substrates as high-performance electrodes for supercapacitors. *Adv. Mater.* **2013**, *25*, 976–979. [CrossRef] [PubMed]
23. Qu, B.H.; Chen, Y.J.; Zhang, M.; Hu, L.L.; Lei, D.N.; Lu, B.G.; Li, Q.H.; Wang, Y.G.; Chen, L.B.; Wang, T.H. β-Cobalt sulfide nanoparticles decorated graphene composite electrodes for high capacity and power supercapacitors. *Nanoscale* **2012**, *4*, 7810–7816. [CrossRef] [PubMed]
24. Yan, J.; Fan, Z.J.; Sun, W.; Ning, G.Q.; Wei, T.; Zhang, Q.; Zhang, R.F.; Zhi, L.J.; Wei, F. Advanced asymmetric supercapacitors based on $Ni(OH)_2$/graphene and porous graphene electrodes with high energy density. *Adv. Funct. Mater.* **2012**, *22*, 2632–2641. [CrossRef]
25. Charles, C.L.M.; Jung, S.; Jonas, C.P.; Thomas, F.J. Benchmarking heterogeneous electrocatalysts for the oxygen evolution reaction. *J. Am. Chem. Soc.* **2013**, *135*, 16977–16987.
26. Qian, H.; Jinlong, L.; Tongxiang, L.; Chen, W. In situ synthesis of mesoporous $NiMoO_4$ on ball milled graphene for high performance supercapacitors. *J. Electrochem. Soc.* **2017**, *164*, E173–E179. [CrossRef]
27. Hatzell, K.B.; Beidaghi, M.; Campos, J.W.; Dennison, C.R.; Kumbur, E.C.; Gogotsi, Y. A high performance pseudocapacitive suspension electrode for the electrochemical flow capacitor. *Electrochim. Acta* **2013**, *111*, 888–897. [CrossRef]
28. Han, D.D.; Xu, P.C.; Jing, X.Y.; Wang, J.; Yang, P.P.; Shen, Q.H.; Liu, J.Y.; Song, D.L.; Gao, Z.; Zhang, M.L. Trisodium citrate assisted synthesis of hierarchical NiO nanospheres with improved supercapacitor performance. *J. Power Sources* **2013**, *235*, 45–53. [CrossRef]
29. Yu, G.H.; Hu, L.B.; Vosgueritchian, M.; Wang, H.L.; Xie, X.; McDonough, J.R.; Cui, X.; Cui, Y.; Bao, Z.N. Solution-processed graphene/MnO_2 nanostructured textiles for high-performance electrochemical capacitors. *Nano Lett.* **2011**, *11*, 2905–2911. [CrossRef] [PubMed]
30. Brousse, T.; B'elanger, D.; Longd, J.W. To be or not to be pseudocapacitive? *J. Electrochem. Soc.* **2015**, *162*, A5185–A5189. [CrossRef]
31. Owusu, K.A.; Qu, L.; Li, J.; Wang, Z.; Zhao, K.; Yang, C.; Hercule, K.M.; Lin, C.; Shi, C.; Wei, Q.; et al. Low-crystalline iron oxide hydroxide nanoparticle anode for high-performance supercapacitors. *Nat. Commun.* **2017**, *8*, 14264. [CrossRef] [PubMed]
32. Liu, M.C.; Kong, L.B.; Lu, C.; Ma, X.J.; Li, X.M.; Luo, Y.C.; Kang, L. Design and synthesis of $CoMoO_4$–$NiMoO_4 \cdot xH_2O$ bundles with improved electrochemical properties for supercapacitors. *J. Mater. Chem. A* **2013**, *1*, 1380–1387. [CrossRef]
33. He, Y.M.; Chen, W.J.; Li, X.D.; Zhang, Z.X.; Fu, J.C.; Zhao, C.H.; Xie, E.Q. Freestanding three-dimensional graphene/MnO_2 composite networks as ultralight and flexible supercapacitor electrodes. *ACS Nano* **2013**, *7*, 174–182. [CrossRef] [PubMed]

34. Prasad, K.R.; Miura, N. Electrochemically deposited nanowhiskers of nickel oxide as a high-power pseudocapacitive electrode. *Appl. Phys. Lett.* **2004**, *85*, 4199–4201. [CrossRef]
35. Brezesinski, T.; Wang, J.; Polleux, J.; Dunn, B.; Tolbert, S.H. Templated nanocrystal-based porous TiO_2 films for next-generation electrochemical capacitors. *J. Am. Chem. Soc.* **2009**, *131*, 1802–1809. [CrossRef] [PubMed]
36. Zheng, J.P.; Jow, T.R. High energy and high power density electrochemical capacitors. *J. Power Sources.* **1996**, *62*, 155–159. [CrossRef]
37. Wang, H.L.; Gao, Q.M.; Jiang, L. Facile approach to prepare nickel cobaltite nanowire materials for supercapacitors. *Small* **2011**, *7*, 2454–2459. [CrossRef] [PubMed]
38. Jiang, H.; Li, C.Z.; Sun, T.; Ma, J. High-performance supercapacitor material based on $Ni(OH)_2$ nanowire-MnO_2 nanoflakes core–shell nanostructures. *Chem. Commun.* **2012**, *48*, 2606–2608. [CrossRef] [PubMed]
39. Ramkumar, R.; Sundaram, M.M. Electrochemical synthesis of polyaniline crosslinked $NiMoO_4$ nanofibre dendrites for energy storage devices. *New J. Chem.* **2016**, *40*, 7456–7464. [CrossRef]
40. Barmi, M.J.; Minakshi, M. Tuning the redox properties of the nanostructured $CoMoO_4$ electrode: effects of surfactant content and synthesis temperature. *ChemPlusChem* **2016**, *81*, 964–977. [CrossRef]
41. Nan, H.; Ma, W.; Gu, Z.; Geng, B.; Zhang, X. Hierarchical $NiMn_2O_4$@CNT nanocomposites for high-performance asymmetric supercapacitors. *RSC Adv.* **2015**, *5*, 24607–24614. [CrossRef]
42. Sugimoto, W.; Iwata, H.; Yasunaga, Y.; Murakami, Y.; Takasu, Y. Preparation of ruthenic acid nanosheets and utilization of its interlayer surface for electrochemical energy storage. *Angew. Chem. Int. Ed.* **2003**, *42*, 4092–4096. [CrossRef] [PubMed]
43. Toupin, M.; Brousse, T.; Be'langer, D. Charge storage mechanism of MnO_2 electrode used in aqueous electrochemical capacitor. *Chem. Mater.* **2004**, *16*, 3184–3190. [CrossRef]

 *nanomaterials* 

Article

Three-Dimensional Honeycomb-Like Porous Carbon with Both Interconnected Hierarchical Porosity and Nitrogen Self-Doping from Cotton Seed Husk for Supercapacitor Electrode

Hui Chen [1], Gang Wang [1,2,3], Long Chen [1], Bin Dai [1] and Feng Yu [1,*]

1 Key Laboratory for Green Processing of Chemical Engineering of Xinjiang Bingtuan, School of Chemistry and Chemical Engineering, Shihezi University, Shihezi 832003, China; huichen09@sina.com (H.C.); wanggang@shzu.edu.cn (G.W.); chenlong323@shzu.edu.cn (L.C.); db_tea@shzu.edu.cn (B.D.)
2 Key Laboratory of Materials-Oriented Chemical Engineering of Xinjiang Uygur Autonomous Region, Shihezi 832003, China
3 Engineering Research Center of Materials-Oriented Chemical Engineering of Xinjiang Production and Construction Corps, Shihezi 832003, China
* Correspondence: yufeng05@mail.ipc.ac.cn; Tel.: +86-993-205-7272; Fax: +86-993-205-7270

Received: 3 May 2018; Accepted: 6 June 2018; Published: 8 June 2018

Abstract: Hierarchical porous structures with surface nitrogen-doped porous carbon are current research topics of interest for high performance supercapacitor electrode materials. Herein, a three-dimensional (3D) honeycomb-like porous carbon with interconnected hierarchical porosity and nitrogen self-doping was synthesized by simple and cost-efficient one-step KOH activation from waste cottonseed husk (a-CSHs). The obtained a-CSHs possessed hierarchical micro-, meso-, and macro-pores, a high specific surface area of 1694.1 m^2/g, 3D architecture, and abundant self N-doping. Owing to these distinct features, a-CSHs delivered high specific capacitances of 238 F/g and 200 F/g at current densities of 0.5 A/g and 20 A/g, respectively, in a 6 mol/L KOH electrolyte, demonstrating good capacitance retention of 84%. The assembled a-CSHs-based symmetric supercapacitor also displayed high specific capacitance of 52 F/g at 0.5 A/g, with an energy density of 10.4 Wh/Kg at 300 W/Kg, and 91% capacitance retention after 5000 cycles at 10 A/g.

Keywords: nitrogen doping; hierarchical structure; porous carbon; three-dimensional architecture; cotton seed husk; supercapacitor; sustainable biomass

---

1. Introduction

Supercapacitors are receiving extensive attention worldwide on account of their fast charge/discharge rates, higher power densities, and excellent cycling stabilities, which are superior to other energy storage devices [1–3]. The energy storage mechanisms of supercapacitors are mainly based on electrical double layer capacitance (EDLC), which occurs at the electrode/electrolyte interface [4,5], and pseudo-capacitors with reversible Faradic redox reaction [6]. Their relatively simple energy storage mechanisms and fast charge/discharge traits make supercapacitors the most promising energy storage devices [7–9]. However, the energy that is stored in a supercapacitor is lower than that in batteries. This has inspired research focused on increasing the supercapacitor's energy density, while maintaining high power density [10].

Recently, studies have been conducted to produce porous carbon-based electrode materials, mainly focusing on porous structures [11,12]. In general, specific surface area (SSA) is the most important feature that influences the electrochemical performances of porous carbon materials [1,13,14]. High SSA benefits include good energy storage, especially for high rate performance. On the other hand, hierarchical pore size distribution accelerates the ion transfer from the electrolyte to electrode surface [15]. Typically, macro- and meso-pores provide space for electrolyte storage, thereby reducing the ion diffusion distance and ion diffusion resistance, which is conducive to better capacitive performance [16–18]. Additionally, three-dimensional (3D) frameworks are capable of supplying stable frame structures, inter-connected pore networks, and are further conducive to ion transfer [18,19]. However, in order to obtain above 3D hierarchical porous carbons (3D-HPCs), it is necessary to add a certain amount of hard templates or soft templates in most of the previously reported works [20–22]. Consequently, the synthesis of 3D-HPCs is limited by its complex, time-consuming, and costly processes.

Surface functional groups also play an important role in energy storage. Nitrogen doping is the most widely studied [23–25]. Generally, a proper amount of nitrogen doping promotes the electrochemical capacitance, by improving the wettability of the porous carbon electrode material, thus bringing about the properties of pseudo-capacitance [26–29]. Nitrogen-doped carbon is achieved by the pyrolysis of nitrogen-enriched polymer precursors and subsequent physicochemical activation, to obtain a nitrogen-doped carbon material. The process is expensive, time-consuming, and endangers our environment [30,31]. On the other hand, through the post-treatment of porous carbon with organic and inorganic nitrogen sources [32], urea [33,34] and ammonia [23,35], it is possible to obtain nitrogen-doped porous carbon. However, the resulting carbon material rarely has the previously mentioned 3D hierarchical structure. The traditional synthesis of 3D hierarchical nitrogen-doped porous carbon (3D-HNPCs) materials is limited to this complex, time-consuming, and costly process. Therefore, transforming sustainable raw materials into highly-performing 3D-HNPC materials, through a simple preparation method, is required.

In this work, we have successfully prepared a 3D hierarchical structure of nitrogen self-doped porous carbon from waste cottonseed husks (CSHs), for high performance supercapacitor electrode materials, using one-step KOH activation. After carbonization and activation, the obtained a-CSHs will have the following features. Firstly, it has a 3D architecture that is associated with hierarchical micro-, meso-, and macro-pores, a high specific surface area of 1694.1 m^2/g, and a moderate pore volume of 0.87 cm^3/g. Secondly, the obtained a-CSHs also contain a moderate nitrogen content of 2.56 atom %, which is as a result of the protein content in the raw materials. Thirdly, the synthesis process is also simple and convenient for large-scale industrial production. The formation of a 3D hierarchical structure is based on the fact that KOH acts both as hard template to form 3D structures and as an activator to produce affluent micropores on the surface of the carbon material. Finally, cotton is an important agricultural crop in China, with an annual production of 1.5 million tons [36]. The Xinjiang Production and Construction Corps occupies up to 70% of the total production. Cottonseed is mainly used to extract cottonseed oil, thereby generating a large amount of sustainable raw material. On the basis of the above advantages, a-CSHs provide high performed high supercapacitor performances in three- and two-electrode systems.

2. Materials and Methods

2.1. Sample Preparation

The preparation of nitrogen self-doped three-dimensional (3D) honeycomb-like porous carbon was synthesized by one-step KOH activation. The specific preparation steps were as follows: Cottonseed husk (CSH) was fully washed with deionized water to remove ash and other impurities, and then dried at 100 °C for 10 h. It was further crushed to form a powder and passed through a 200 mesh sieve for further use. Subsequently, the CSH powder was vigorously stirred with an aqueous KOH solution at a mass ratio of KOH/CSH powder = 1, and then dried at 80 °C. The mixture was then activated in a tube furnace under an Ar atmosphere at 600 °C, 700 °C, or 800 °C for 1 h with a heating rate of 5 °C/min. The obtained product was washed with 10% v/v HCl to remove the metal impurities, was washed with deionized water until the pH of the filtrate was 7.0, and was dried at 80 °C for 10 h. The obtained samples were denoted a-CSH-x, where x represented the activation temperature.

2.2. Material Characterization

Scanning electron microscopy (SEM) surveys were examined with a Hitachi SU8010 microscope (Tokyo, Japan). Transmission electron microscopy (TEM) and energy dispersive X-ray spectroscopy (EDS) were analyzed by a field emission Tecnai G2 F20 electron (Hillsboro, OR, USA) microscope. X-ray diffraction (XRD) measurements were carried out on a Bruker D8 Advance X-ray diffractometer with Cu-Kα radiation (Karlsruhe, Germany). Specific surface areas of the samples were calculated using the Brunauer-Emmett-Teller (BET) method (Micromeritics ASAP 2020 BET apparatus, Atlanta, GA, USA). The pore size distribution (PSD) curves were derived from the adsorption branch, using a nonlocal density functional theory (NLDFT) model assuming slit pore geometry. The surface chemical compositions were determined using an ESCALAB 250Xi (Thermo Fisher Scientific, USA) X-ray photoelectron spectroscope (XPS). The Raman spectra were collected on a LabRAM HR800 Laser Confocal Micro-Raman Spectroscope (Horiba Jobin Yvon, Franch) with a laser wavelength of 532 nm.

2.3. Electrochemical Measurements

The electrochemical properties of the as-prepared samples were tested on a CHI 760E working station with a 6 M KOH electrolyte. Cyclic voltammetry (CV) tests at different scanning rates and galvanostatic charge/discharge (GCD) curves under varying current densities were used to evaluate the electrochemical performances of the electrode materials. The working electrode was obtained by mixing carbon material (5 mg) with acetylene black (1 mg) and polytetrafluoroethylene (1 μL) in absolute ethanol (1 mL). The mixture was dispersed by ultrasound for 40 min and the ink-like dispersion that was obtained was transferred to nickel foam (1 cm × 1 cm) and then vacuum dried at 80 °C for 10 h. The nickel foam was further pressed on a tablet press at 20 MPa for 1 min and was immersed in 6 M KOH for further testing. The loaded mass of each electrode was 5 mg. For the three-electrode system, the Pt sheet and Saturated Calomel Electrode (SCE) were utilized as counter electrode and reference electrode, respectively. The specific capacitances of the samples were calculated through discharge curves following Equation (1), as follows:

$$C = \frac{I \times \Delta t}{m \times \Delta V} \tag{1}$$

where C (F/g) is the specific capacitance, I (A) is the charge/discharge current, Δt (s) is the discharging time, m (g) is the mass of the working electrode, and ΔV (v) is the voltage window of the charge/discharge process.

The electrochemical properties of a-CSH-700 were measured with the two electrode system. Two symmetrical electrodes were separated by a cellulose membrane in a 6 M KOH electrolyte and were assembled in a CR2032 stainless-steel coin cell. The specific capacitance was calculated from the discharge process, according to Equation (1). The energy density and power density of symmetric supercapacitor systems were further calculated by Equations (2) and (3).

$$E = \frac{1}{2} C_t \Delta V^2 \times \frac{1}{3.6} \tag{2}$$

$$P = \frac{E}{\Delta t} \times 3600 \tag{3}$$

where E (Wh/kg), P (W/kg), C_t (F/g), ΔV (v), and Δt (h) are the specific energy density, specific power density, specific capacitance, and voltage window, respectively, of the symmetrical supercapacitor system.

3. Results

Nitrogen self-doped 3D honeycomb-like porous carbon was prepared through one-step activation from waste cottonseed husk (CSH). Cotton seeds were generally collected for the preparation of cottonseed oil; large amounts of cotton seed husks were not effectively used and were abandoned. Therefore, we recycled them to prepare high-performance biomass-derived electrode materials (Figure 1a,b). The material preparation process is shown in Figure 1b–d. Pretreated CSH powder (Figure 1c) was directly stirred with aqueous KOH solution and dried for carbonization and activation. The temperature for activation was adjusted from 600 °C to 800 °C, and the carbon that was obtained was washed and dried. The entire preparation process was cost-efficient, simple, and easily achieved the industrialized requirements.

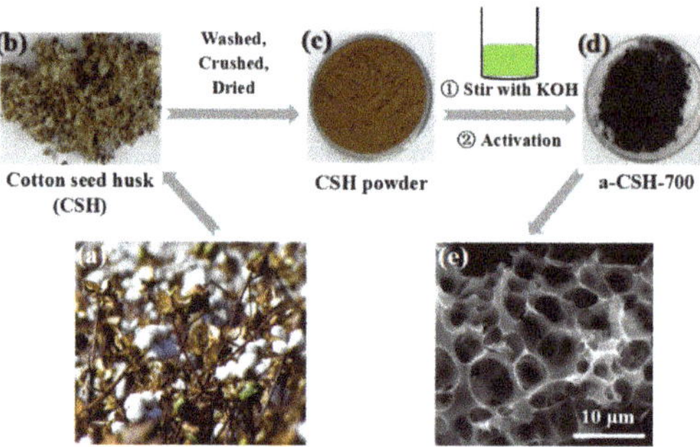

Figure 1. Schematic diagrams for the fabrication of a-CSH-700 and the corresponding scanning electron microscopy (SEM) image. (**a**) Images of cotton. (**b–d**) Schematic of the synthesis of a-CSH-700 derived from cottonseed husk, and (**e**) the corresponding SEM images of a-CSH-700.

Scanning electron microscopy (SEM) images of a-CSH-600, 700, and 800 are shown in Figure 2a–f. The micromorphologies of the obtained samples showed typical 3D inter-connected honeycomb-like microstructures at different pyrolysis temperatures. The chemical composition of the waste cottonseed husk had a certain degree of degradation after stirring and evaporation with an aqueous KOH solution. Subsequently, the following chemical reaction of CSH and KOH during carbonization and activation processes occurred: $6KOH + C \rightarrow 2K + 3H_2 + 2K_2CO_3$, followed by the decomposition of K_2CO_3, and the simultaneous generation of 3D pore structures and graphite sheet-like layer structures [8,37,38]. The lateral size of the 3D porous carbon varied in the range 400 nm to 4 μm (Figure 1d–f). With the increasing pyrolysis temperature, the characteristics of the 3D structure were slightly damaged, which was mainly because the higher temperature was bad for obtaining the 3D structure. The 3D linked carbon skeleton caused the obtained carbon material to exhibit a higher specific surface area. Moreover, this unique 3D structure generated abundant interconnected pore structure that allowed the electrolyte to be stored therein, reducing the distance that the electrolyte travelled on the surface of the electrode material. At the same time, abundant micropores and mesoporous structures existed on the surface of the carbon material (Figure 3a,b). With the help of these multi-level pore structures, it was easy to obtain a high energy storage performance.

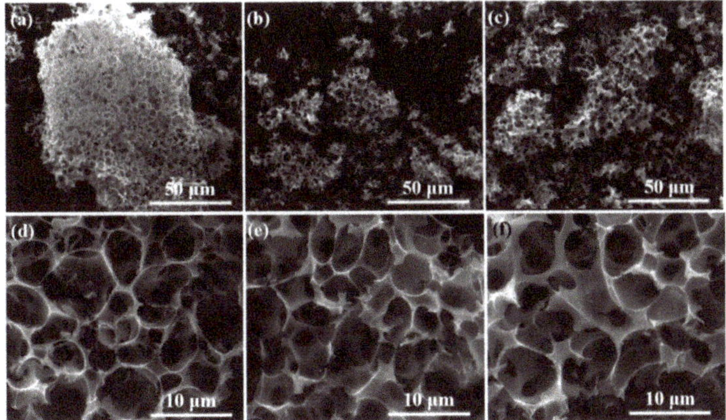

Figure 2. SEM images of (**a**,**d**) a-CSH-600, (**b**,**e**) a-CSH-700, and (**c**,**f**) a-CSH-800.

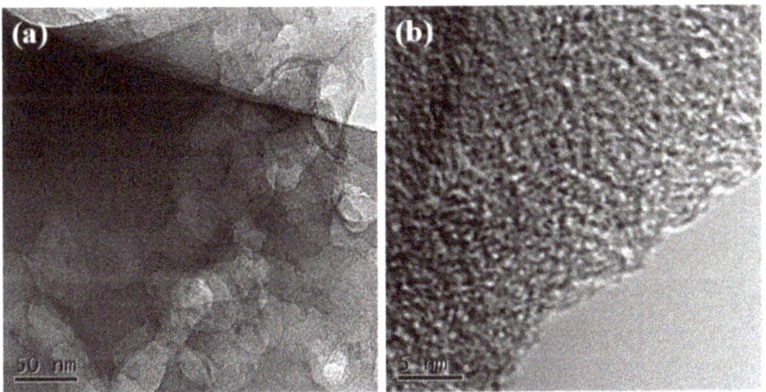

Figure 3. Transmission electron microscopy (TEM) image (**a**) and high-resolution TEM (HRTEM) image (**b**) of a-CSH-700.

The powder X-ray diffraction (XRD) patterns of as-prepared a-CSHs samples are shown in Figure 4a. There were two distinct peaks at around 2θ = 22.1° and 43.5°, which were ascribed to the (002) and (100) reflections of the amorphous graphitic carbon structure. The high intensity values at the low angles indicated high specific surface areas of carbon materials. The obvious peak at 43.5° revealed a higher degree of interlayer condensation in a-CSHs, which also significantly increased the electrical conductivity. Raman spectroscopy was further used to characterize the a-CSHs samples. As shown in Figure 4b, there were three distinct peaks at 1343 cm^{-1} (D band), 1590 cm^{-1} (G band), and 2800 cm^{-1} (2D band). The D band represented the degree of the defects and disordered sp^3 carbon atoms in the sample, and the G band was in line with graphite sp^2 hybridized carbon atoms in the sample. The presence of a 2D peak indicated that there existed an ordered graphite-like structure in the a-CSHs. The intensity ratio of D band to G band (I_D/I_G) represented the disorder degree of the samples [39,40]. The I_D/I_G ratios of a-CSH-600, a-CSH-700, and a-CSH-800 were 0.83, 0.84, and 0.87, respectively. We confirmed that the chemical reaction became deeper with rise in activation temperature, which promoted the defects and disordered structures in a-CSH-700 and a-CSH-800.

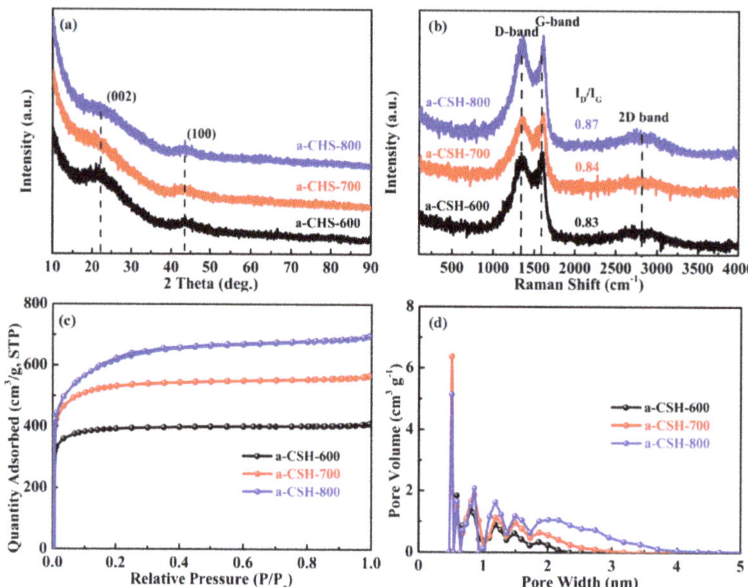

Figure 4. Powder X-ray diffraction (XRD) patterns (**a**) and Raman spectra (**b**) of a-CSHs samples. (**c**) Nitrogen adsorption-desorption isotherms and (**d**) pore size distributions of a-CSHs.

The nitrogen adsorption/desorption measurements were further used to examine the pore properties of a-CSHs. The nitrogen adsorption-desorption isotherm and pore size distribution curve of a-CSHs are shown in Figure 4c,d, respectively. It could be seen that all of the a-CSHs samples displayed type I isotherms [41]. With the increase in pyrolysis temperature, the corresponding quantity adsorbed value also increased [42]. This suggested that the specific surface area increased with an increase in the activation temperature. Figure 4d displays the pore size distribution isotherms. The dominant pore size distribution was located in the micropores (0.5–2 nm), and a part in mesopores (2–4 nm). a-CSH-600, a-CSH-700, and a-CSH-800 exhibited hierarchical porous structures, abundant micropores, and profuse mesopores, respectively, which were consistent with the results of the SEM images. Table 1 also summarizes the information on the specific BET surface areas and pore sizes of all of the a-CSHs samples. The specific surface areas of a-CSH-600, a-CSH-700, and a-CSH-800 were determined to be 1257.8, 1694.1, and 2063.0 m^2/g, respectively, while the pore volumes were 0.64, 0.87, and 1.07 cm^3/g, respectively. This showed that the temperature of activation was a dominating factor for the development of pore structure.

Table 1. Pore characteristics of the a-CSHs samples.

Samples	S_{BET} [a] (m^2/g)	S_{mi} [b] (m^2/g)	V_{total} [c] (cm^3/g)	V_{mid} [b] (m^2/g)	D_{aver} [d] (nm)
a-CSH-600	1257.8	1051.0	0.64	0.51	3.95
a-CSH-700	1694.1	1253.5	0.87	0.63	3.76
a-CSH-800	2063.0	1080.0	1.07	0.52	2.60

[a] Total surface area calculated using the Brunauer-Emmett-Teller (BET) method; [b] Micropore surface area and volume calculated from the t-plot method; [c] Total pore volume calculated at $P/P_o = 0.99$; [d] Average pore diameter calculated from the (Barrett-Joyner-Halenda) BJH desorption.

The chemical compositions and surface functional groups of as-prepared a-CSH-600, a-CSH-700, and a-CSH-800 were further characterized by XPS measurements. As shown in the survey spectra

(Figure 5a), there existed three distinct peak signals corresponding to the C 1s peak near 286 eV, the N 1s peak near 400 eV, and the O 1s peak near 534 eV. Table 2 lists the statistical results of the corresponding elemental contents of the as-prepared carbon materials. The carbon content decreased and the associated oxygen content increased with rise in activation temperature. The nitrogen content of a-CSHs ranged from 1.51 to 2.56 atom %, the main reason being the complex chemical processes and structure features of these groups at a higher temperature. Figure 6 shows the element mapping images of as-obtained a-CSH-700. The nitrogen was uniformly distributed on the surface of the carbon, and the nitrogen content was 2.62 atom %.

Figure 5b–d displays high resolution C 1s, O 1s, and N 1s spectra of the as-prepared carbon materials. The high resolution C 1s spectrum (Figure 5b) of a-CSHs was resolved into four individual peaks at 284.6, 285.7, 286.7, and 289.2 eV, corresponding to C=C, C–C, C–O, and C=O, respectively [43,44]. The high-resolution O 1s spectrum at 530.8, 531.9, 533, and 534.1 eV were associated with COOH, O–C=O/N–C=O, C–O=C, and C–OH/N–O–C, respectively [45,46]. The high-resolution N 1s spectrum was composed of pyridinic-N (398.4 eV), pyrolic-N (400.1 eV), quaternary-N (401 eV), and oxidized-N (402.7 eV) [47–49]. These nitrogen and oxygen functional groups enhanced the surface wettabilities of the a-CSHs, and thereby promoting the specific capacitances of the as-prepared carbon materials [8,50,51].

Table 2. Elemental contents of a-CSHs samples from X-ray photoelectron spectroscope (XPS) and EDS.

Samples	Composition (from XPS)			(from EDX)
	C (atom %)	O (atom %)	N (atom %)	N (atom %)
a-CSH-600	90.03	8.46	1.51	1.87
a-CSH-700	90.11	7.33	2.56	2.62
a-CSH-800	91.19	6.37	2.44	2.86

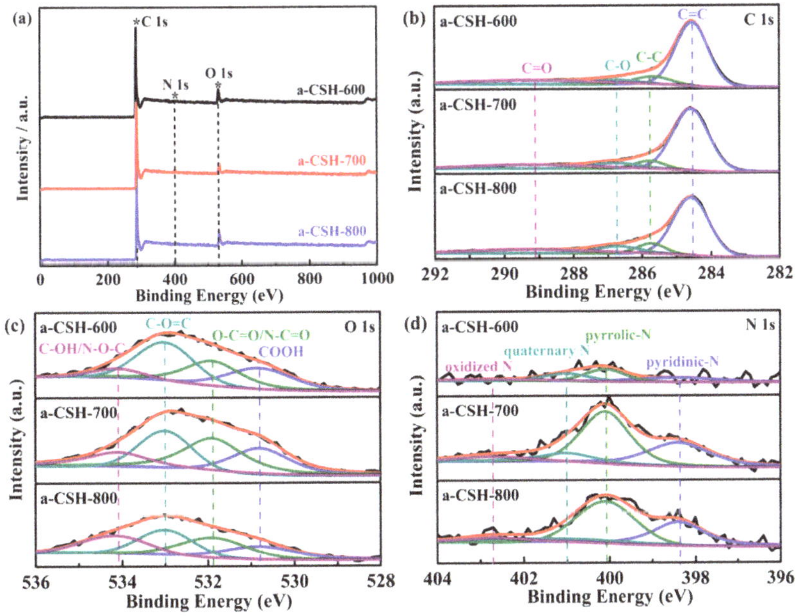

Figure 5. XPS images of the a-CSH-600, a-CSH-700, and a-CSH-800 (**a**). High-resolution C 1s (**b**), O 1s (**c**), and N 1s (**d**) of the a-CSH-600, a-CSH-700, and a-CSH-800.

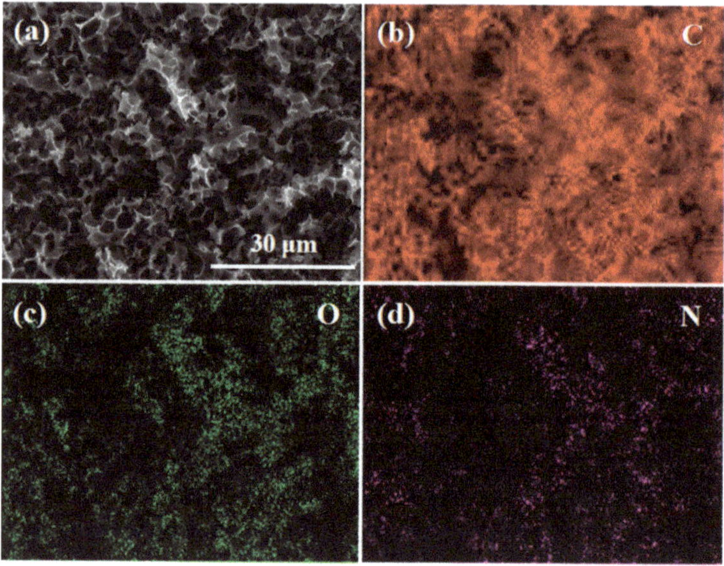

Figure 6. (**a**) SEM image of a-CSH-700, and elemental mapping images of (**b**) C, (**c**) O, and (**d**) N.

The general standards for designing high performance supercapacitor electrodes were associated with high specific capacitance, good rate capability, and long cycle stability. According to the above conclusions, the a-CSHs derived from one-step synthesis possessed many advantages for supercapacitor electrode materials. The a-CSHs had superior specific surface area for forming ideal electrochemical double layers. The interconnected micropores and mesopores could increase ion transport and the abundant surface nitrogen functional groups improved the wettability and promoted electrical conductivity. The 3D structure provided space for electrolyte storage. These characteristics endowed a-CSHs with a high supercapacitor performance.

The electrochemical performances of the obtained a-CSHs were tested by a three-electrode system in a 6 M KOH aqueous electrolyte and the CV curves are shown in Figure 7. The CV curves of a-CSH-600 (Figure 7a), a-CSH-700 (Figure 7b), and a-CSH-800 (Figure 7c) showed typical quasi-rectangular shapes at scan rates from 5 mV/s to 50 mV/s. This suggested that all of the samples displayed ideal electrochemical double layer capacitances. The CV curves of a-CSH-700 had the largest current responses and areas, suggesting the highest capacitance. Furthermore, galvanostatic charge/discharge (GCD) was further used to assess the electrochemical performance, which is shown in Figure 8. The GCD curves of a-CSH-600 (Figure 8a), a-CSH-700 (Figure 8b), and a-CSH-800 (Figure 8c) showed typical quasi-linear shapes at current densities from 0.5 A/g to 20 A/g. It also suggested that the as-obtained carbon samples possessed good electrochemical performances. Figure 8d summarizes the gravimetric specific capacitances of a-CSHs, calculated from galvanostatic charge/discharge curves at a current density ranging from 0.5 to 20 A/g. The specific capacitance value of the a-CSH-700 sample was as high as 238 F/g at a current density of 0.5 A/g, which was higher than those of a-CSH-800 (217 F/g) and a-CSH-600 (204 F/g). The capacitance was also as high as 200 F/g for a-CSH-700 with an excellent capacitance retention of 92%, indicating its outstanding rate capability. For a better comparison, Table 3 lists the specific capacitances of other biomass-derived porous carbons that were reported in the recent literature. Although the electrode of a-CSH-700 exhibited the best electrochemical performance, its specific area and surface functional groups were not the highest among all of the samples, meaning that there were synergistic effects.

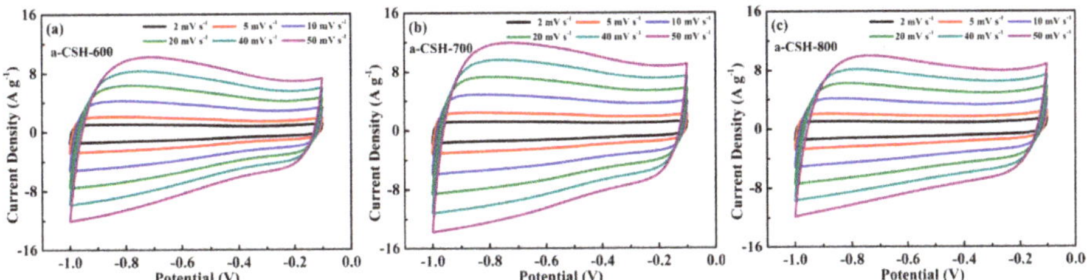

Figure 7. Electrochemical performance characteristics of a-CSHs measured in a three-electrode system in a 6 M KOH electrolyte: the cyclic voltammetry (CV) curves of (**a**) a-CSH-600, (**b**) a-CSH-700, and (**c**) a-CSH-800 at different scan rates.

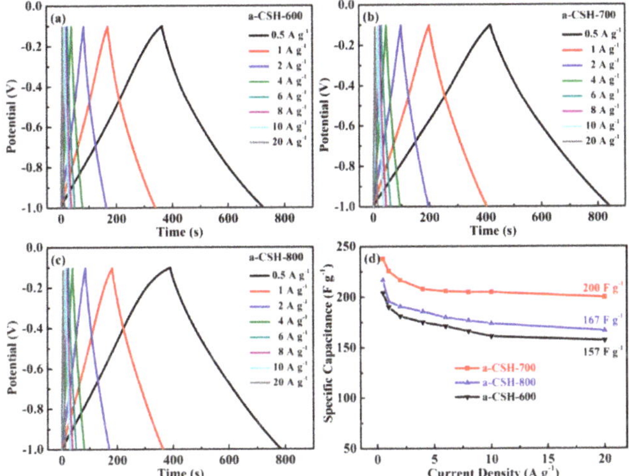

Figure 8. Electrochemical performance characteristics of a-CSHs measured in a three-electrode system in a 6 M KOH electrolyte: galvanostatic charge/discharge (GCD) curves of (**a**) a-CSH-600, (**b**) a-CSH-700, and (**c**) a-CSH-800 at different scan rates. (**d**) Specific capacitances at different current densities.

Table 3. Electrochemical performance of biomass derived porous carbons.

Materials	S_{BET} (m^2/g)	C_m (F/g)	Current Density	Electrolyte	Ref.
Broad beans	655.4	129	10 A/g	6 M KOH	[52]
Banana peel	1357.6	155	2.5 A/g	6 M KOH	[44]
Potato waste	1052	192	10 A/g	2 M KOH	[53]
Banana peel	1650	182	10 A/g	6 M KOH	[54]
Sugarcane bagasse	1939.6	175	20 A/g	1 M H_2SO_4	[55]
Pomelo	974.6	176.4	20 A/g	2 M KOH	[56]
Chitin	1600	196.2	20 A/g	6 M KOH	[57]
Cotton seed husk	1694.1	200	20 A/g	6 M KOH	This work

The capacitive performance of a-CSH-700 was further evaluated by assembling it in a symmetric two-electrode cell in 6 M KOH. Figure 9a shows the CV curves that were tested in different potential windows. It could be seen that there no obvious promotion of anodic current when the operating voltage was 1.2 V. The GCD curves (Figure 9b) had a symmetric triangular shape as the current density increased from 0.5 A/g to 10 A/g, and there were no evident IR drops. The testing results also showed that it had a good capacitive performance. The specific capacitance for the entire electrochemical supercapacitor was estimated to be 52 F/g at 0.5 A/g with an energy density of 10.4 Wh/kg and power density 300 W/kg (Figure 9c,e). Figure 9d shows the cycling stability of the device at a higher current density of 10 A/g. It could be seen that the charge/discharge curves of the last two cycles were the same as the first two cycles (Figure 9f). Capacitance retention was also as high as 91% after 5000 cycles. These results revealed that the synthesized a-CSH-700 sample was a superior electrode material for high power and cost-effective supercapacitors.

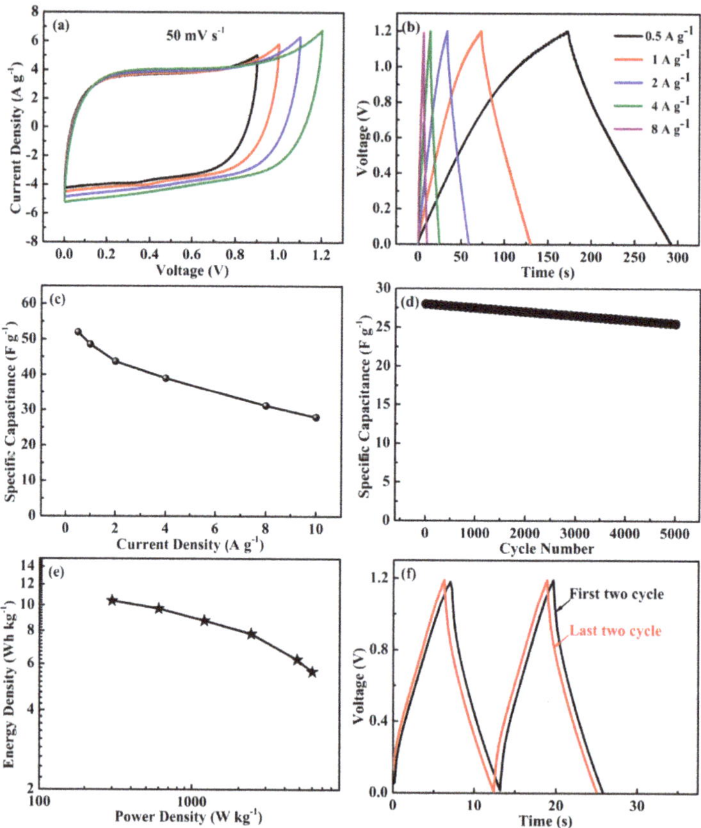

Figure 9. Electrochemical measurements of an as-assembled a-CSH-700//a-CSH-700 symmetric supercapacitor in a 6 M KOH electrolyte: (**a**) CV curves of the cell operated in different voltage windows at scan rate 50 mV/s; (**b**) galvanostatic charge/discharge curves of the cell at various current densities; and (**c**) specific capacitances for the supercapacitor at different current densities. (**d**) Cycling stabilities of the devices at a current density of 10 A/g. (**e**) Ragone plot. (**f**) First two cycle charge-discharge and last two charge-dischage plots.

4. Conclusions

Nitrogen self-doped 3D honeycomb-like porous carbon was successfully prepared via one-step KOH activation (a-CSHs). When the carbonization temperature was adjusted, the obtained carbon materials showed high specific surface areas (1257.8–2063.0 m^2/g), abundant nitrogen contents (1.51–2.56 atom %), and hierarchical pore structures. The a-CSH-700 based electrode delivered a high specific capacitance of 238 F/g and 200 F/g at current densities of 0.5 A/g and 20 A/g, respectively, demonstrating a good capacitance retention of 84%. Moreover, the assembled a-CSH-700 based symmetric supercapacitor also displayed a high specific capacitance of 52 F/g at 0.5 A/g, with an energy density of 10.4 Wh/kg at 300 W/kg, and 91% capacitance retention after 5000 cycles at 10 A/g. This facile and cost-effective method for the preparation of 3D honeycomb-like porous carbon material from waste cottonseed husk was also renewable and easy for industrial production.

Author Contributions: F.Y. designed and administered the experiments. H.C. performed the experiments; H.C., G.W., L.C., and B.D. characterized the samples and collected the data; and F.Y., G.W., and B.D. gave technical support and conceptual advice. All of the authors contributed to the analysis and discussion of the data and to writing the manuscript.

Funding: This research was funded by the National Natural Science Foundation of China (Grant No. U1303291), the Program for Changjiang Scholars and Innovative Research Team in University (Grant No. IRT_15R46), and the Program of Science and Technology Innovation Team in Bingtuan (No. 2015BD003).

Conflicts of Interest: The authors declare no conflict of interest.

References

1. Zhang, L.L.; Zhao, X.S. Carbon-based materials as supercapacitor electrodes. *Chem. Soc. Rev.* **2009**, *38*, 2520–2532. [CrossRef] [PubMed]
2. Miller, J.R.; Simon, P. Electrochemical capacitors for energy management. *Science* **2008**, *321*, 651–652. [CrossRef] [PubMed]
3. Qin, T.; Wan, Z.; Wang, Z.; Wen, Y.; Liu, M.; Peng, S.; He, D.; Hou, J.; Huang, F.; Cao, G. 3D flexible O/N co-doped graphene foams for supercapacitor electrodes with high volumetric and areal capacitances. *J. Power Sources* **2016**, *336*, 455–464. [CrossRef]
4. Zhao, G.; Chen, C.; Yu, D.; Sun, L.; Yang, C.; Zhang, H.; Sun, Y.; Besenbacher, F.; Yu, M. One-step production of O-N-S co-doped three-dimensional hierarchical porous carbons for high-performance supercapacitors. *Nano Energy* **2018**, *47*, 547–555. [CrossRef]
5. Qin, T.; Peng, S.; Hao, J.; Wen, Y.; Wang, Z.; Wang, X.; He, D.; Zhang, J.; Hou, J.; Cao, G. Flexible and wearable all-solid-state supercapacitors with ultrahigh energy density based on a carbon fiber fabric electrode. *Adv. Energy Mater.* **2017**, *7*. [CrossRef]
6. Wen, Y.; Peng, S.; Wang, Z.; Hao, J.; Qin, T.; Lu, S.; Zhang, J.; He, D.; Fan, X.; Cao, G. Facile synthesis of ultrathin $NiCo_2S_4$ nano-petals inspired by blooming buds for high-performance supercapacitors. *J. Mater. Chem. A* **2017**, *5*, 7144–7152. [CrossRef]
7. Liu, L.; Niu, Z.; Chen, J. Unconventional supercapacitors from nanocarbon based electrode materials to device configurations. *Chem. Soc. Rev.* **2016**, *45*, 4340–4363. [CrossRef] [PubMed]
8. Wang, J.; Kaskel, S. Koh activation of carbon-based materials for energy storage. *J. Mater. Chem.* **2012**, *22*, 23710–23725. [CrossRef]
9. Zhang, Y.; Liu, X.; Wang, S.; Li, L.; Dou, S. Bio-nanotechnology in high-performance supercapacitors. *Adv. Energy Mater.* **2017**, *7*. [CrossRef]
10. Yan, J.; Wang, Q.; Wei, T.; Fan, Z. Recent advances in design and fabrication of electrochemical supercapacitors with high energy densities. *Adv. Energy Mater.* **2014**, *4*. [CrossRef]
11. Liu, L.; Niu, Z.; Chen, J. Flexible supercapacitors based on carbon nanotubes. *Chinese Chem. Lett.* **2018**, *4*, 571–581. [CrossRef]
12. Hou, J.; Cao, C.; Idrees, F.; Ma, X. Hierarchical porous nitrogen-doped carbon nanosheets derived from silk for ultrahigh-capacity battery anodes and supercapacitors. *ACS Nano* **2015**, *9*, 2556–2564. [CrossRef] [PubMed]
13. Raymundo-Piñero, E.; Kierzek, K.; Machnikowski, J.; Béguin, F. Relationship between the nanoporous texture of activated carbons and their capacitance properties in different electrolytes. *Carbon* **2006**, *44*, 2498–2507. [CrossRef]
14. Chen, T.; Dai, L. Carbon nanomaterials for high-performance supercapacitors. *Mater. Today* **2013**, *16*, 272–280. [CrossRef]
15. Wu, X.; Jiang, L.; Long, C.; Fan, Z. From flour to honeycomb-like carbon foam: Carbon makes room for high energy density supercapacitors. *Nano Energy* **2015**, *13*, 527–536. [CrossRef]

16. Liu, Y.; Xiao, Z.; Liu, Y.; Fan, L.Z. Biowaste-derived 3d honeycomb-like porous carbon with binary-heteroatom doping for high-performance flexible solid-state supercapacitors. *J. Mater. Chem. A* **2017**, *6*, 160–166. [CrossRef]
17. Salanne, M.; Rotenberg, B.; Naoi, K.; Kaneko, K.; Taberna, P.L.; Grey, C.P.; Dunn, B.; Simon, P. Efficient storage mechanisms for building better supercapacitors. *Nat. Energy* **2017**, *1*, 16070. [CrossRef]
18. Forse, A.C.; Merlet, C.; Griffin, J.M.; Grey, C.P. Newperspectives on the charging mechanisms of supercapacitors. *J. Am. Chem. Soc.* **2016**, *138*, 5731–5744. [CrossRef] [PubMed]
19. Yu, Z.; Tetard, L.; Zhai, L.; Thomas, J. Supercapacitor electrode materials: Nanostructures from 0 to 3 dimensions. *Energy Environ. Sci.* **2015**, *8*, 702–730. [CrossRef]
20. Dutta, S.; Bhaumik, A.; Wu, C.W. Hierarchically porous carbon derived from polymers and biomass: Effect of interconnected pores on energy applications. *Energy Environ. Sci.* **2014**, *7*, 3574–3592. [CrossRef]
21. Wang, X.; Zhang, Y.; Zhi, C.; Wang, X.; Tang, D.; Xu, Y.; Weng, Q.; Jiang, X.; Mitome, M.; Golberg, D. Three-dimensional strutted graphene grown by substrate-free sugar blowing for high-power-density supercapacitors. *Nat. Commun.* **2013**, *4*, 2905. [CrossRef] [PubMed]
22. Ghimire, P.; Gunathilake, C.; Wickramaratne, N.P.; Jaroniec, M. Tetraethyl orthosilicate-assisted synthesis of nitrogen-containing porous carbon spheres. *Carbon* **2017**, *121*, 408–417. [CrossRef]
23. Li, B.; Dai, F.; Xiao, Q.; Yang, L.; Shen, J.; Zhang, C.; Cai, M. Nitrogen-doped activated carbon for high energy hybrid supercapacitor. *Energy Environ. Sci.* **2015**, *9*, 102–106. [CrossRef]
24. Yang, W.; Hou, L.; Xu, X.; Li, Z.; Ma, X.; Yang, F.; Li, Y. Carbon nitride template-directed fabrication of nitrogen-rich porous graphene-like carbon for high performance supercapacitors. *Carbon* **2018**. [CrossRef]
25. Wang, C.; Wang, F.; Liu, Z.; Zhao, Y.; Liu, Y.; Yue, Q.; Zhu, H.; Deng, Y.; Wu, Y.; Zhao, D. N-doped carbon hollow microspheres for metal-free quasi-solid-state full sodium-ion capacitors. *Nano Energy* **2017**, *41*, 674–680. [CrossRef]
26. Wei, L.; Sevilla, M.; Fuertes, A.B.; Mokaya, R.; Yushin, G. Hydrothermal carbonization of abundant renewable natural organic chemicals for high-performance supercapacitor electrodes. *Adv. Energy Mater.* **2011**, *1*, 356–361. [CrossRef]
27. Wei, X.; Zou, H.; Gao, S. Chemical crosslinking engineered nitrogen-doped carbon aerogels from polyaniline-boric-acid-polyvinyl-alcohol gels for high-performance electrochemical capacitors. *Carbon* **2017**, *123*, 471–480. [CrossRef]
28. Wang, C.; Zhou, Y.; Sun, L.; Wan, P.; Zhang, X.; Qiu, J. Sustainable synthesis of phosphorus- and nitrogen-co-doped porous carbons with tunable surface properties for supercapacitors. *J. Power Sources* **2013**, *239*, 81–88. [CrossRef]
29. Jeong, H.M.; Lee, J.W.; Shin, W.H.; Choi, Y.J.; Shin, H.J.; Kang, J.K.; Choi, J.W. Nitrogen-doped graphene for high-performance ultracapacitors and the importance of nitrogen-doped sites at basal planes. *Nano Lett.* **2011**, *11*, 2472–2477. [CrossRef] [PubMed]
30. Su, F.; Poh, C.K.; Chen, J.S.; Xu, G.; Wang, D.; Li, Q.; Lin, J.; Lou, X.W. Nitrogen-containing microporous carbon nanospheres with improved capacitive properties. *Energy Environ. Sci.* **2011**, *4*, 717–724. [CrossRef]
31. Yan, J.; Wei, T.; Qiao, W.; Fan, Z.; Zhang, L.; Li, T.; Zhao, Q. A high-performance carbon derived from polyaniline for supercapacitors. *Electrochem. Commun.* **2010**, *12*, 1279–1282. [CrossRef]
32. Ornelas, O.; Sieben, J.M.; Ruizrosas, R.; Morallón, E.; Cazorlaamorós, D.; Geng, J.; Soin, N.; Siores, E.; Johnson, B.F. On the origin of the high capacitance of nitrogen-containing carbon nanotubes in acidic and alkaline electrolytes. *Chem. Commun.* **2014**, *50*, 11343–11346. [CrossRef] [PubMed]
33. Zou, K.; Deng, Y.; Chen, J.; Qian, Y.; Yang, Y.; Li, Y.; Chen, G. Hierarchically porous nitrogen-doped carbon derived from the activation of agriculture waste by potassium hydroxide and urea for high-performance supercapacitors. *J. Power Sources* **2018**, *378*, 579–588. [CrossRef]
34. Tian, W.; Zhang, H.; Sun, H.; Tadé, M.O.; Wang, S. Template-free synthesis of n-doped carbon with pillared-layered pores as bifunctional materials for supercapacitor and environmental applications. *Carbon* **2017**, *118*, 98–105. [CrossRef]
35. Kim, N.D.; Kim, W.; Ji, B.J.; Oh, S.; Kim, P.; Kim, Y.; Yi, J. Electrochemical capacitor performance of n-doped mesoporous carbons prepared by ammoxidation. *J. Power Sources* **2008**, *180*, 671–675. [CrossRef]
36. Chen, M.; Kang, X.; Dou, J.; Gao, B.; Han, Y.; Xu, G.; Liu, Z.; Zhang, L. Preparation of activated carbon from cotton stalk and its application in supercapacitor. *J. Solid State Electrochem.* **2013**, *17*, 1005–1012. [CrossRef]

37. Zhang, Q.; Han, K.; Li, S.; Li, M.; Li, J.; Ren, K. Synthesis of garlic skin-derived 3D hierarchical porous carbon for high-performance supercapacitors. *Nanoscale* **2018**, *10*, 2427. [CrossRef] [PubMed]
38. Qian, W.; Sun, F.; Xu, Y.; Qiu, L.; Liu, C.; Wang, S.; Yan, F. Human hair-derived carbon flakes for electrochemical supercapacitors. *Energy Environ. Sci.* **2013**, *7*, 379–386. [CrossRef]
39. Gao, S.; Geng, K.; Liu, H.; Wei, X.; Zhang, M.; Wang, P.; Wang, J. Transforming organic-rich amaranthus waste into nitrogen-doped carbon with superior performance of oxygen reduction reaction. *Energy Environ. Sci.* **2014**, *8*, 221–229. [CrossRef]
40. Chen, H.; Guo, Y.C.; Wang, F.; Wang, G.; Qi, P.R.; Guo, X.H.; Dai, B.; Yu, F. An activated carbon derived from tobacco waste for use as a supercapacitor electrode material. *New Carbon Mater.* **2017**, *32*, 592–599. [CrossRef]
41. Kai, W.; Ning, Z.; Lei, S.; Rui, Y.; Tian, X.; Wang, J.; Yan, S.; Xu, D.; Guo, Q.; Lang, L. Promising biomass-based activated carbons derived from willow catkins for high performance supercapacitors. *Electrochim. Acta* **2015**, *166*, 1–11.
42. Xie, L.; Sun, G.; Su, F.; Guo, X.; Kong, Q.Q.; Li, X.M.; Huang, X.; Wan, L.; Song, W.; Li, K. Hierarchical porous carbon microtubes derived from willow catkins for supercapacitor application. *J. Mater. Chem. A* **2015**, *4*, 1637–1646. [CrossRef]
43. Kim, N.D.; Buchholz, D.B.; Casillas, G.; José-Yacaman, M.; Chang, R.P.H. Hierarchical design for fabricating cost-effective high performance supercapacitors. *Adv. Funct. Mater.* **2014**, *24*, 4186–4194. [CrossRef]
44. Liu, B.; Zhang, L.; Qi, P.; Zhu, M.; Wang, G.; Ma, Y.; Guo, X.; Chen, H.; Zhang, B.; Zhao, Z. Nitrogen-doped banana peel-derived porous carbon foam as binder-free electrode for supercapacitors. *Nanomaterials* **2016**, *6*, 18. [CrossRef] [PubMed]
45. Huang, Y.; Peng, L.; Liu, Y.; Zhao, G.; Chen, J.Y.; Yu, G. Biobased nano porous active carbon fibers for high-performance supercapacitors. *ACS Appl. Mater. Interfaces* **2016**, *8*, 15205–15215. [CrossRef] [PubMed]
46. Wang, Y.; Zhu, M.; Wang, G.; Dai, B.; Yu, F.; Tian, Z.; Guo, X. Enhanced oxygen reduction reaction by in situ anchoring Fe_2N nanoparticles on nitrogen-doped pomelo peel-derived carbon. *Nanomaterials* **2017**, *7*, 404. [CrossRef] [PubMed]
47. Shi, L.; Wu, T.; Wang, Y.; Zhang, J.; Wang, G.; Zhang, J.; Dai, B.; Yu, F. Nitrogen-doped carbon nanoparticles for oxygen reduction prepared via a crushing method involving a high shear mixer. *Materials* **2017**, *10*, 1030. [CrossRef] [PubMed]
48. Wang, Y.; Zhu, M.; Li, Y.; Zhang, M.; Xue, X.; Shi, Y.; Dai, B.; Guo, X.; Yu, F. Heteroatom-doped porous carbon from methyl orange dye wastewater for oxygen reduction. *Green Energy Environ.* **2017**. [CrossRef]
49. Wang, Y.; Yu, F.; Zhu, M.; Ma, C.; Zhao, D.; Wang, C.; Zhou, A.; Dai, B.; Ji, J.; Guo, X. N-doping of plasma exfoliated graphene oxide via dielectric barrier discharge plasma treatment for oxygen reduction reaction. *J. Mater. Chem. A* **2018**, *6*, 2011–2017. [CrossRef]
50. Ania, C.O.; Khomenko, V.; Raymundo-Piñero, E.; Parra, J.B.; Béguin, F. The large electrochemical capacitance of microporous doped carbon obtained by using a zeolite template. *Adv. Funct. Mater.* **2007**, *17*, 1828–1836. [CrossRef]
51. Wu, J.; Zhang, D.; Wang, Y.; Hou, B. Electrocatalytic activity of nitrogen-doped graphene synthesized via a one-pot hydrothermal process towards oxygen reduction reaction. *J. Power Sources* **2013**, *227*, 185–190. [CrossRef]
52. Xu, G.Y.; Han, J.P.; Bing, D.; Ping, N.; Jin, P.; Hui, D.; Li, H.S.; Zhang, X.G. Biomass-derived porous carbon materials with sulfur and nitrogen dual-doping for energy storage. *Green Chem.* **2015**, *17*, 1668–1674. [CrossRef]
53. Ma, G.; Yang, Q.; Sun, K.; Peng, H.; Ran, F.; Zhao, X.; Lei, Z. Nitrogen-doped porous carbon derived from biomass waste for high-performance supercapacitor. *Bioresour. Technol.* **2015**, *197*, 137–142. [CrossRef] [PubMed]
54. Lv, Y.; Gan, L.; Liu, M.; Xiong, W.; Xu, Z.; Zhu, D.; Wright, D.S. A self-template synthesis of hierarchical porous carbon foams based on banana peel for supercapacitor electrodes. *J. Power Sources* **2012**, *209*, 152–157. [CrossRef]
55. Wang, B.; Wang, Y.; Peng, Y.; Wang, X.; Wang, J.; Zhao, J. 3-dimensional interconnected framework of n-doped porous carbon based on sugarcane bagasse for application in supercapacitors and lithium ion batteries. *J. Power Sources* **2018**, *390*, 186–196. [CrossRef]

56. Peng, H.; Ma, G.; Sun, K.; Zhang, Z.; Yang, Q.; Lei, Z. Nitrogen-doped interconnected carbon nanosheets from pomelo mesocarps for high performance supercapacitors. *Electrochim. Acta* **2016**, *190*, 862–871. [CrossRef]
57. Jie, Z.; Li, B.; Wu, S.; Wei, Y.; Hui, W. Chitin based heteroatom-doped porous carbon as electrode materials for supercapacitors. *Carbohydr. Polym.* **2017**, *173*, 321–329.

Letter

High Electrochemical Performance Phosphorus-Oxide Modified Graphene Electrode for Redox Supercapacitors Prepared by One-Step Electrochemical Exfoliation

Lei Zu [1], Xing Gao [1,2], Huiqin Lian [1], Xiaomin Cai [1,2], Ce Li [1,2], Ying Zhong [1], Yicheng Hao [1], Yifan Zhang [1], Zheng Gong [1], Yang Liu [1], Xiaodong Wang [2] and Xiuguo Cui [1,*]

[1] School of Material Science and Engineering, Beijing Institute of Petrochemical Technology, Beijing 102617, China; zulei@bipt.edu.cn (L.Z.); gaoxing@bipt.edu.cn (X.G); lianhuiqin@bipt.edu.cn (H.L.); 13522433853@163.com (X.C.); lice0520@163.com (C.L.); zhongying@bipt.edu.cn (Y.Z.); haoyicheng@bipt.edu.cn (Y.H.); zhangyifan@bipt.edu.cn (Y.Z.); gongzheng@bipt.edu.cn (Z.G); yang.liu@bipt.edu.cn (Y.L.)

[2] State Key Laboratory of Organic-Inorganic Composites, Beijing University of Chemical Technology, Beijing 100029, China; wangxd@mail.buct.edu.cn

* Correspondence: cuixiuguo@bipt.edu.cn; Tel.: +86-010-81294007

Received: 18 May 2018; Accepted: 4 June 2018; Published: 9 June 2018

Abstract: Phosphorus oxide modified graphene was prepared by one-step electrochemical anodic exfoliation method and utilized as electrode in a redox supercapacitor that contained potassium iodide in electrolytes. The whole preparation process was completed in a few minutes and the yield was about 37.2%. The prepared sample has better electrocatalysis activity for I^-/I^-_3 redox reaction than graphite due to the good charge transfer performance between phosphorus oxide and iodide ions. The maximum discharge specific capacitance is 1634.2 F/g when the current density is 3.5 mA/cm^2 and it can keep at 463 F/g after 500 charging–discharging cycles when the current density increased about three times.

Keywords: graphene; electrochemical anodic exfoliation; phosphorus oxide group; supercapacitors; iodide ions

1. Introduction

In recent years, graphene and its composite materials have attracted great attention due to their excellent performance in various fields. Many researchers have presented that graphene plays an important role as electrodes, catalysts, and supports in energy-storage fields, such as supercapacitors, ions batteries, fuel cells and so on [1–4]. To improve the electrochemical performance of graphene, introducing other active groups (–OH, –COOH, –NO$_x$, and –PO$_x$) or heteroatoms, including O, P, N and B, into graphene is regarded as an effective way to adjust the electrochemical properties of graphene. The advantages of this strategy are the heteroatoms could tailor the electronic properties and average electronic charge distribution of adjacent carbon atoms of graphene. Among various heteroatom-doped graphene materials, nitrogen-doped (N-doped) graphene has been widely studied in supercapacitors [5] and batteries [6]. However, only a few studies are reported so far regarding graphene modified with phosphorus (P) and phosphorus-oxide (–PO$_x$) in electrochemical energy storage field. Recently, Karthika et al. prepared a high electrochemical activity P-doped graphene by reducing graphene oxide with phosphoric acid. The prepared sample exhibited a specific discharge capacitance value of 367 F/g, which is much larger than that of graphene [7]. This study suggests that the electrochemical properties of graphene can be effectively improved by introducing P into graphene. Therefore, it is very necessary to study the preparation and electrochemical properties of P and P–O group modified graphene. In addition, it was revealed that P-modified graphene was much more stable in air than N-modified graphene and exhibited improved n-type semiconducting behavior [8]. However, the traditional preparation methods are relatively complex and time-consuming. Wang et al. synthesized P/C hybrid through a simple ball-milling approach under argon protection. The whole preparation time was 16 h [9]. Xu et al. prepared P doped graphene via thermally annealing approach and the preparation time was more than 48 h [10]. To overcome these drawbacks, it is necessary to use another fast and efficiency method to prepare P modified graphene material. In contrast to the tradition methods, the electrochemical exfoliation approach is believed to be an effective and efficiency strategy. The whole time of preparation usually takes only a few minutes. Accordingly, the preparation of P or P–O group modified graphene through electrochemical approach is more advantageous than traditional methods.

Currently, supercapacitors (SCs) play an important role in the field of energy storage and have been applied in many areas due to high power density, good rate capability and excellent cycling stability [11]. Different from the traditional methods, a unique strategy of introducing redox additives in the electrolytes could effective improve the capacity of SCs via redox reactions at the interface of electrode and electrolyte. Therefore, the efficient use of redox electrolytes is an indispensable way to increase the electrochemical performance of SCs. In our previous studies, we have presented that the potassium iodide (KI) is an effective active redox electrolyte to enhance the capacitance of SCs due to the redox reaction between pairs of $3I^-/I^-_3$, $2I^-/I_2$, $2I^-_3/I_2$, I_2/IO^-_3, etc. [12–19]. The electrode materials include conductive polymers such as polyaniline and poly(3,4-ethylenedioxythiophene) (PEDOT):poly(styrene sulfonate) (PSS). Although these polymers have good electrochemical performance, the preparation process is tedious, and, because of the stability restriction of polymer material itself, iodide ions may destroy the structure when they frequently penetrate and release from the polymer. Therefore, it is necessary to apply a new material with good electrochemical performance and that is not easily damaged to improve the performance of the supercapacitor.

In recent studies, some researchers have presented that heteroatom-doped graphene has excellent electrocatalytic activity for the reduction of iodine, due to the electrons conjugation effect, charge redistribution and defect sites [20]. However, the electrochemical catalytic activity of phosphorus oxide modified graphene (PO-graphene) has not been fully revealed. Similar to doping heteroatoms, introducing P–O groups is an effective way to improve the electrochemical performance of graphene. The advantages lie in the modification of the electronic properties and the effect on electronic charge distribution of graphene. Furthermore, the P–O groups could also perform redox reactions with some electroactive ions, such as iodide ions. Therefore, the PO-graphene is a suitable electrode material to use in supercapacitors that contains KI electrolysis.

In this paper, we present a facile method to prepare high quality PO-graphene and study the electrochemical property. The PO-graphene is prepared by the one-step electrochemical anodic exfoliation method and used as the electrode in a redox supercapacitor, whose electrolyte contains KI. We investigated in detail the catalytic effect of P–O group on redox reaction of $3I^-/I^-_3$ via various electrochemical analysis. We present that the introduced P–O group in graphene could effectively improve its electrochemical performance. The whole preparation process can be completed within 30 min and the yield is about 37.2%. Owing to the catalytic effect of P–O group, the PO-graphene has a better electrochemical performance than graphite in promoting the redox reaction of I^-/I^-_3.

2. Materials and Methods

2.1. Materials

High purity graphite rod was obtained from C Nano Technology Co. Ltd., Beijing, China. Ammonium phosphate, Poly(vinylidene fluoride) (PVDF), Sulfuric acid (H_2SO_4) and potassium iodide (KI) were obtained from Aladdin Co. Ltd., Shanghai, China.

2.2. Phosphorus Oxide Modified Graphene (PO-Graphene) Preparation

In a typical procedure, high purity graphite rod and platinum plate ($1 \times 1 \times 0.05$ cm^3) were used as the anode and cathode, respectively, and the distance was fixed at 1.5 cm. The exfoliation was performed at 10 V in 1 M ammonium phosphate aqueous. After the preparation, the sample was collected by centrifugation at 5000 rpm for 30 min and purified repeatedly with distilled water until PH = 7.

2.3. Characterization

The as-prepared PO-graphene was characterized by field emission scanning electron microscope (FESEM; Zeiss EVO-50, Zeiss, Germany), Transmission electron microscopy (TEM) images are performed on a Tecnal G2 20STWIN microscope operating (FEI, Hillsboro, OR, USA) at 200 kV. Fourier transform Infrared spectroscopy (FT-IR) analyses were accomplished on Thermal Nicolet 6700 (Thermo Nicolet, Shanghai, China). Laser Raman spectra were carried out using Raman spectrometry (RENISHAW inVia Raman Microscope, London, UK). X-ray diffraction analysis (XRD) was carried out by Bruker D8 Advance (BRUKER AXS, Berlin, Germany), Cu Kα = 0.154 nm with a voltage of 40 kV.

The electrochemical measurements were performed by a three-electrode cell. In detail, the working electrode was composed of PO-graphene (90 wt %) and polyvinylidene fluoride PVDF (10 wt %). A Pt foil electrode and a saturated calomel electrode (SCE) (CH Instruments Ins, Shanghai, China) were used as the counter electrode and reference electrode, respectively. The electrolytes were composed of 0.1 M KI and 0.2 M H_2SO_4. The cyclic voltammetry (CV) tests were performed from −0.5 to 0.7 V (vs. SCE). Electrochemical impedance spectroscopy (EIS) measurements were performed under open circuit conditions over a frequency region from 0.01 Hz to 100 kHz by applying an AC signal of 5 mV in amplitude throughout the test. The galvanostatic charge–discharge (GCD) analyses were performed in a potential range of 0–0.7 V (vs. SCE). The CV, GCD and EIS tests were all performed on a CHI660D electrochemical workstation (CH Instruments Ins, Shanghai, China).

3. Results

The schematic diagram of the preparation and the photo of PO-graphene are shown in Figure 1a,b. The preparation process included two parts: oxidation and exfoliation of graphite. The oxidation process produced the P–O groups and then the graphite was exfoliated into few-layer graphene, which possessed a platelet-like morphology (Figure 1c,d), with the promotion of water decomposition. The whole preparation process was completed within 30 min and the yield reached 37.2%.

The chemical composition of PO-graphene was determined by FT-IR, XPS and Raman spectra. As presented in Figure 2a, the PO-graphene shows clear stretching peaks of P=O, O–P–O, P–OH, and P-H at 1155, 1116, 946, and 2443–2330 cm^{-1}, respectively. Different types of oxygen functionalities are presented at 3434 cm^{-1} (O–H stretching), 1076 cm^{-1} (C–O stretching), 980 and 861 cm^{-1} (P–O–C stretching). The peaks at 1649 (C=C skeletal vibrations from graphite) and 2856 cm^{-1} (CH_2 stretching) are also detected [21]. These characteristic peaks confirm the graphene is successfully modified by P–O group.

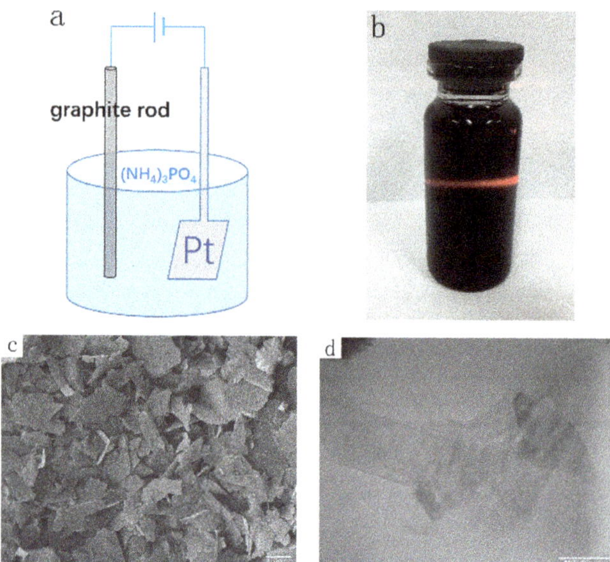

Figure 1. (**a**) schematic diagram of preparation; (**b**) the photo of phosphorus oxide modified graphene (PO-graphene); (**c**) scanning electron microscope (SEM) and (**d**) transmission electron microscopy (TEM) image of PO-graphene.

The high resolution XPS spectra could provide detailed bonding and chemical information for the PO-graphene. As presented in Figure 2b, the C 1s peak can be divided into three components centered at about 284.3, 285.8 and 287.5 eV, which cloud be attributed to C–C, C–P, and C–O bonding, respectively [22]. The presence of the C–P characteristic peak confirms that P atoms have been successfully bonded into the graphene. The high-resolution P 2p spectrum (Figure 2c) shows that phosphorus bonded with graphene in two types of chemical bonding: P–C and P–O bonding, located at about 129.8 and 133.7 eV, respectively [23]. The presence of both P–C and P–O stretch models confirms that the P–O group was successfully bonded with graphene.

The Raman spectra of PO-graphene and graphite are presented in Figure 2d. The spectrum of the original graphite shows a very weak D band at 1338 cm^{-1} and strong G band at 1563 cm^{-1}. The D band is related to the defects and disorder in the structure of the graphite due to the intervalley scattering, while the G band corresponds to the E_{2g} vibration of sp^2 of C atoms. In contrast to the origin graphite, the PO-graphene shows a clearly strong D band. The intensity ratio of D to G band (I_D/I_G) increases from 0.1 to 0.87, manifesting that the disorder degree and structural defect (decrease in the average size of the sp^2 domains) in PO-graphene are enhanced due to the oxidization during the preparation [24]. The decreased layers of graphene in PO-graphene were also demonstrated by the wave shift of the G band from 1563 to 1568 cm^{-1}.

The XRD patterns of PO-graphene and graphite are presented in Figure 2e. PO-graphene exhibits two diffraction peaks centered at 23.7° and 26.3°, corresponding to the larger interlayer spacing of 0.375 and 0.338 nm, respectively. Compared with the graphite (about 0.335 nm), the larger interlayer distance is due to the accommodation of oxygen functional groups and water molecules during the preparation due to the obvious hydrophilic character [25]. This result also suggests the formation of P–O groups on graphene.

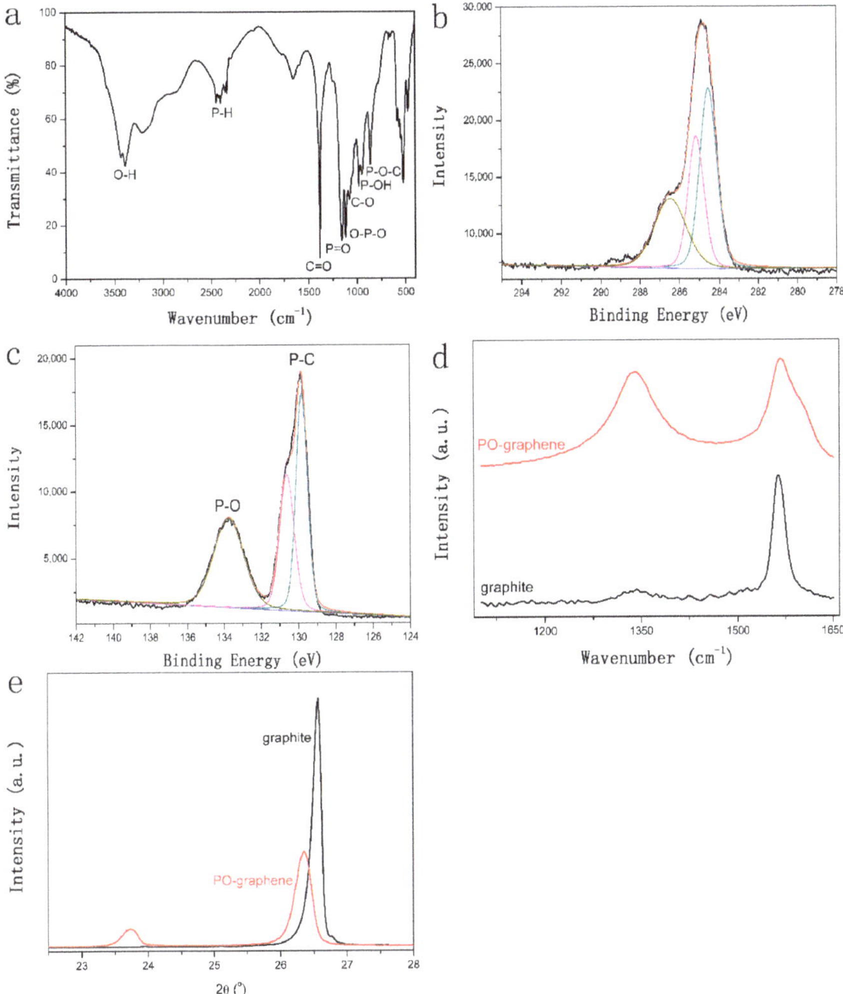

Figure 2. (**a**) Fourier transform Infrared spectroscopy (FT-IR) analysis of PO-graphene; (**b**) high-resolution C1s spectrum of PO-graphene; (**c**) high-resolution P2p spectrum of PO-graphene; (**d**) Raman analysis patterns of PO-graphene and graphite and (**e**) X-ray diffraction analysis (XRD) patterns of PO-graphene and graphite.

The electrochemical performance of PO-graphene is investigated by CV analysis. As shown in Figure 3a, compared with the graphite, which possesses only one single reduction peak, the PO-graphene has a more symmetry and larger peak current than that of graphite, suggesting that it has a better electrochemical performance due to the catalysis of P–O groups. The pseudo capacitance performance comes from both the redox transformation of $I^- \leftrightarrow I^-{}_3$ and the interaction between P–O groups and iodide ions. Clearly, the PO-graphene possesses three pair of redox peaks. The peaks centered at 0.27 and 0.12 V are due to the transformation of $3I^- - 2e \leftrightarrow I^-{}_3$, the peaks at 0.53 and 0.22 V could be assigned to the transformation of $5I^- - 4e \leftrightarrow I^-{}_5$ and the peaks at 0.65 and 0.29 V may be attributed to the interaction between P–O and iodide ions.

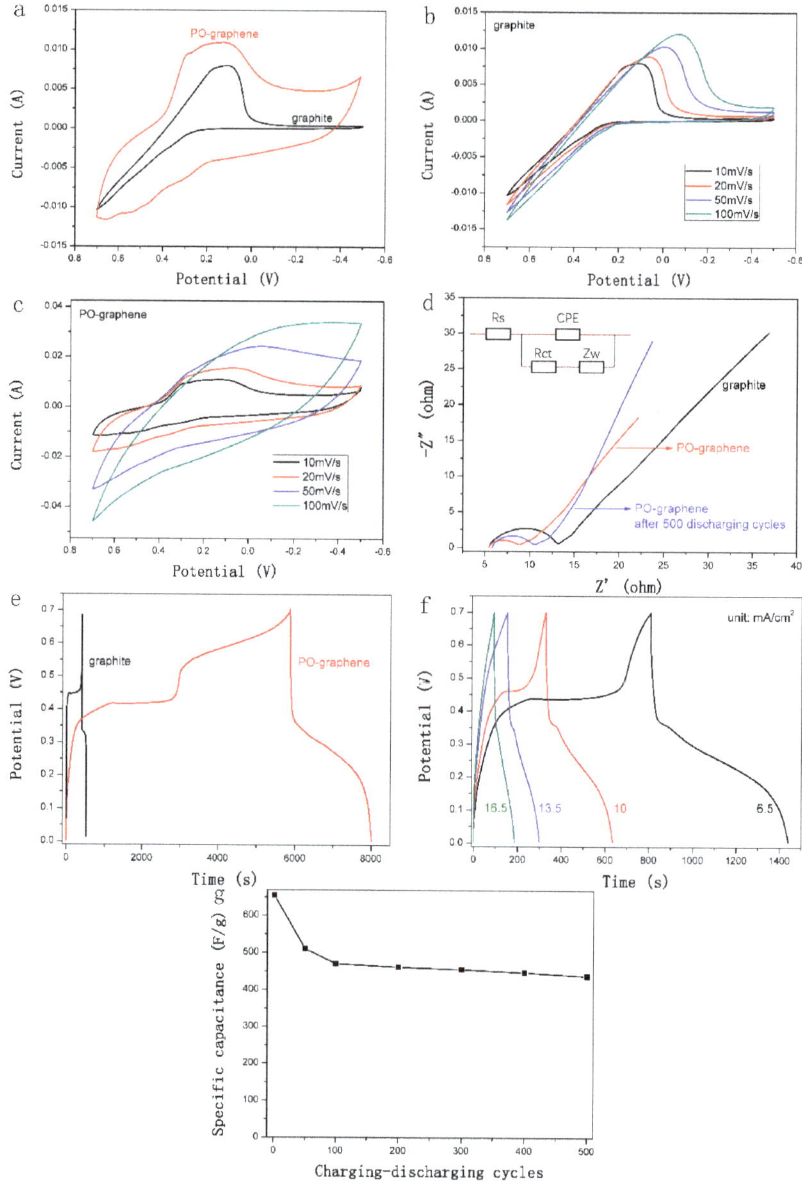

Figure 3. (**a**) Cyclic voltammetry (CV) profiles of PO-graphene and graphite performed in a mixture of 0.1 M KI and 0.2 M H_2SO_4 aqueous at 10 mV/s; (**b**) CV profiles of PO-graphene and (**c**) graphite at different scan rate; (**d**) Nyquist plots and equivalent circuit (inset) of PO-graphene and graphite; (**e**) galvanostatic charge–discharge (GCD) analysis of PO-graphene and graphite at 3.5 mA/cm^2; and (**f**) GCD analysis of PO-graphene at different current density; (**g**) cycling stability analysis at 10 mA/cm^2.

To investigate the charge transfer process between PO-graphene and iodide ions, CV tests at different scan rates were performed. As exhibited in Figure 3b, with increasing scan rate, the separation of the peak potentials was enlarged slightly, indicating that the oxidation and reduction of iodide ions in graphite electrode were surface-controlled process. In contrast, the redox peak current of PO-graphene (Figure 3c) was increased apparently by improving the scan rate and the shape of the curve gradually turns into a rectangle, suggesting that the interaction between PO-graphene and iodide ions was performed through a Faradic reaction.

The charge transfer performance is also investigated by EIS analysis. The Nyquist plots and equivalent circuit are shown in Figure 3d. In the Nyquist plots, the diameter of the semicircle manifests the difficulty of the charge transfer in the electrochemical system [26]. As can be seen clearly, the PO-graphene shows a smaller semicircle than that of the graphite, suggesting that it has a lower charge transfer resistance and a faster charge transfer process. The charge-transfer resistance for PO-graphene and graphite is 2.09 and 6.67 Ω/cm^2, respectively. This result could be ascribed to the strong redox interaction between P–O groups and iodide ions, and this conclusion agrees well with the CV results.

Because of the convenient charge transfer character, the PO-graphene has a better charging–discharging performance than that of graphite. As shown in Figure 3e, the discharge specific capacitance of PO-graphene and graphite is calculated as 1634.2 and 87.1 F/g, respectively. The rate capacity of PO-graphene was tested with different current density. As presented in Figure 3f, the discharge specific capacitance decreased gradually with increasing current density from 6.5 mA/cm^2 to 16.5 mA/cm^2 due to the polarization. The maximum discharge specific capacitance was calculated as 904.3 F/g when the current density was 6.5 mA/cm^2 and it reached 342.5 F/g at 16.5 mA/cm^2. This result manifests that the PO-graphene possesses a good rate performance and this discharge performance is also better than the other electrochemical systems, as shown in Table 1. The cycling stability of PO-graphene was investigated at the current density of 10 mA/cm^2. As shown in Figure 3g, the discharge specific capacitance still keeps 438 F/g after 500 charge–discharge cycles. This good cycling stability is ascribed to the excellent charge transfer ability of PO-graphene. The charge-transfer resistance only increased 3.1 Ω/cm^2 compared to pristine during the charge–discharge process (as presented in Figure 3d, inset).

Table 1. Comparison of discharge specific capacitance between the phosphorus oxide modified graphene (PO-graphene) and other reported values.

Materials	Electrolyte	Test Condition	Specific Capacitance (F/g)	Reference
PX-MWCNT	H_2SO_4, 1 M	20 mA/cm^2	118	[27]
PANI/SWCNT	H_2SO_4, 1 M	5 mA/cm^2	185	[28]
PANI/MWCNT	$NaNO_3$, 1 M	5 mA/cm^2	328	[29]
PANI/MWCNT	H_2SO_4, 0.5 M	5 mA/cm^2	500	[15]
N-rGO	H_2SO_4, 1 M	1 mA/cm^2	459	[30]
PEDOT-GF	H_2SO_4, 1 M	2 mA/cm^2	522	[31]
PO-graphene	KI, 0.1 M + H_2SO_4, 0.2 M	3.5 mA/cm^2 6.5 mA/cm^2 10 mA/cm^2 13.5 mA/cm^2 16.5 mA/cm^2	1634.2 904.3 655.7 437.1 342.5	Present work

4. Conclusions

In summary, we applied electrochemical anodic exfoliation method to prepare phosphorus oxide modified graphene and used it as an effective electrode in a redox supercapacitor. The PO-graphene has excellent electrochemical performance due to the significant catalysis to iodide ions and convenient charge transfer capability between phosphorus oxide and iodide ions. The maximum discharge capacitance is 1634.2 F/g when the current density is 3.5 mA/cm^2. The PO-graphene possesses good rate performance and cycling stability.

Author Contributions: L.Z., H.L. and X.C. conceived and designed the experiments; L.Z., H.L., X.G., C.L., X.C., Y.Z., Y.H., Y.Z. and Z.G. performed the experiments and analyzed the data; Y.L., X.W. and X.C. contributed the analysis tools; and L.Z. and H.L. wrote the paper.

Acknowledgments: This work was supported by the Natural Science Foundation of China (NSFC, Nos. 51573021, 21271031, 51063009, and 51203012); the Beijing Natural Science Foundation of China (Nos. 2132009 and 2122015); Innovation Promotion Project of Beijing Municipal Commission of Education, China (No. TJSHG2015 11021002); Opening Foundation of the State Key Laboratory of Organic-Inorganic Composites (oic-201801009); and Talents project of Beijing Organization Department (No. 2017000020124G095).

Conflicts of Interest: The authors declare no conflict of interest.

References

1. Dai, L. Functionalization of graphene for efficient energy conversion and storage. *Acc. Chem. Res.* **2013**, *46*, 31–42. [CrossRef] [PubMed]
2. Li, H.; Sun, L.; Wang, Z.; Zhang, Y.; Tan, T.; Wang, G.; Bakenov, Z. Three-dimensionally hierarchical graphene based aerogel encapsulated sulfur as cathode for lithium/sulfur batteries. *Nanomaterials* **2018**, *8*, 69–80. [CrossRef] [PubMed]
3. Dhanabalan, A.; Li, X.; Agrawal, R.; Chen, C.; Wang, C. Fabrication and characterization of SnO_2/graphene composites as high capacity anodes for Li-Ion batteries. *Nanomaterials* **2013**, *3*, 606–614. [CrossRef] [PubMed]
4. Huang, G.; Kang, W.; Geng, Q.; Xing, B.; Liu, Q.; Jia, J.; Zhang, C. One-step green hydrothermal synthesis of few-layer graphene oxide from humic acid. *Nanomaterials* **2018**, *8*, 215–223. [CrossRef] [PubMed]
5. Yang, Z.; Chuangang, H.; Yue, H.; Huhu, C.; Gaoquan, S.; Liangti, Q. A versatile, ultralight, nitrogen-doped graphene framework. *Angew. Chem. Int. Ed.* **2012**, *51*, 11371–11375.
6. Reddy, A.L.M.; Srivastava, A.; Gowda, S.R.; Gullapalli, H.; Dubey, M.; Ajayan, P.M. Synthesis of nitrogen-doped graphene films for lithium battery application. *ACS Nano* **2010**, *4*, 6337–6342. [CrossRef] [PubMed]
7. Karthika, P.; Rajalakshmi, N.; Dhathathreyan, K.S. Phosphorus-doped exfoliated graphene for supercapacitor electrodes. *J. Nanosci. Nanotechnol.* **2013**, *13*, 1746–1751. [CrossRef] [PubMed]
8. Surajit, S.; Jangah, K.; Keunsik, L.; Atul, K.; Yeoheung, Y.; SaeMi, L.; Taesung, K.; Hyoyoung, L. Highly air-stable phosphorus-doped n-type graphene field-effect transistors. *Adv. Mater.* **2012**, *24*, 5481–5486.
9. Song, J.; Yu, Z.; Gordin, M.L.; Hu, S.; Yi, R.; Tang, D.; Walter, T.; Regula, M.; Choi, D.; Li, X. Chemically bonded phosphorus/graphene hybrid as a high performance anode for sodium-ion batteries. *Nano Lett.* **2014**, *14*, 6329–6335. [CrossRef] [PubMed]
10. Yu, Z.; Song, J.; Gordin, M.L.; Yi, R.; Tang, D.; Wang, D. Phosphorus-graphene nanosheet hybrids as lithium-ion anode with exceptional high-temperature cycling stability. *Adv. Sci.* **2015**, *2*, 1400020–1400026. [CrossRef] [PubMed]
11. Han, Y.; Ge, Y.; Chao, Y.; Wang, C.; Wallace, G.G. Recent progress in 2D materials for flexible supercapacitors. *J. Energy Chem.* **2018**, *27*, 57–72. [CrossRef]
12. Meng, F.; Li, Q.; Zheng, L. Flexible fiber-shaped supercapacitors: Design, fabrication, and multi-functionalities. *Energy Storage Mater.* **2017**, *8*, 85–109. [CrossRef]
13. Zhang, Y.; Zu, L.; Lian, H.; Hu, Z.; Jiang, Y.; Liu, Y.; Wang, X.; Cui, X. An ultrahigh performance supercapacitors based on simultaneous redox in both electrode and electrolyte. *J. Alloys Compd.* **2017**, *694*, 136–144. [CrossRef]
14. Hu, Z.; Zu, L.; Jiang, Y.; Lian, H.; Liu, Y.; Wang, X.; Cui, X. High performance nanocomposite electrodes of mesoporous silica platelet-polyaniline synthesized via impregnation polymerization. *Polym. Compos.* **2017**, *38*, 1616–1623. [CrossRef]
15. Zhang, J.; Kong, L.-B.; Wang, B.; Luo, Y.-C.; Kang, L. In-situ electrochemical polymerization of multi-walled carbon nanotube/polyaniline composite films for electrochemical supercapacitors. *Synth. Met.* **2009**, *159*, 260–266. [CrossRef]
16. Zu, L.; Cui, X.; Jiang, Y.; Hu, Z.; Lian, H.; Liu, Y.; Jin, Y.; Li, Y.; Wang, X. Preparation and electrochemical characterization of mesoporous polyaniline-silica nanocomposites as an electrode material for pseudocapacitors. *Materials* **2015**, *8*, 1369–1383. [CrossRef] [PubMed]
17. Zhang, Y.; Cui, X.; Zu, L.; Cai, X.; Liu, Y.; Wang, X.; Lian, H. New supercapacitors based on the synergetic redox effect between electrode and electrolyte. *Materials* **2016**, *9*, 734–746. [CrossRef] [PubMed]

18. Hu, Z.; Zu, L.; Jiang, Y.; Lian, H.; Liu, Y.; Li, Z.; Chen, F.; Wang, X.; Cui, X. High specific capacitance of polyaniline/mesoporous manganese dioxide composite using KI-H_2SO_4 electrolyte. *Polymers* **2015**, *7*, 1939–1953. [CrossRef]
19. Gao, X.; Zu, L.; Cai, X.; Li, C.; Lian, H.; Liu, Y.; Wang, X.; Cui, X. High performance of supercapacitor from pedot:Pss electrode and redox iodide ion electrolyte. *Nanomaterials* **2018**, *8*, 335–347. [CrossRef] [PubMed]
20. Senthilkumar, S.T.; Selvan, R.K.; Melo, J.S. Redox additive/active electrolytes: A novel approach to enhance the performance of supercapacitors. *J. Mater. Chem. A* **2013**, *1*, 12386–12394. [CrossRef]
21. Chen, M.; Shao, L.-L.; Guo, Y.-X.; Cao, X.-Q. Nitrogen and phosphorus co-doped carbon nanosheets as efficient counter electrodes of dye-sensitized solar cells. *Chem. Eng. J.* **2016**, *304*, 303–312. [CrossRef]
22. Choi, C.H.; Park, S.H.; Woo, S.I. Phosphorus-nitrogen dual doped carbon as an effective catalyst for oxygen reduction reaction in acidic media: Effects of the amount of p-doping on the physical and electrochemical properties of carbon. *J. Mater. Chem.* **2012**, *22*, 12107–12115. [CrossRef]
23. Yang, D.-S.; Bhattacharjya, D.; Inamdar, S.; Park, J.; Yu, J.-S. Phosphorus-doped ordered mesoporous carbons with different lengths as efficient metal-free electrocatalysts for oxygen reduction reaction in alkaline media. *J. Am. Chem. Soc.* **2012**, *134*, 16127–16130. [CrossRef] [PubMed]
24. Wang, Z.; Li, P.; Chen, Y.; He, J.; Liu, J.; Zhang, W.; Li, Y. Phosphorus-doped reduced graphene oxide as an electrocatalyst counter electrode in dye-sensitized solar cells. *J. Power Sour.* **2014**, *263*, 246–251. [CrossRef]
25. Siow, K.S.; Britcher, L.; Kumar, S.; Griesser, H.J. Deposition and xps and ftir analysis of plasma polymer coatings containing phosphorus. *Plasma Process. Polym.* **2014**, *11*, 133–141. [CrossRef]
26. Sykam, N.; Mohan Rao, G. Room temperature synthesis of reduced graphene oxide nanosheets as anode material for supercapacitors. *Mater. Lett.* **2017**, *204*, 169–172. [CrossRef]
27. Potphode, D.D.; Sivaraman, P.; Mishra, S.P.; Patri, M. Polyaniline/partially exfoliated multi-walled carbon nanotubes based nanocomposites for supercapacitors. *Electrochim. Acta* **2015**, *155*, 402–410. [CrossRef]
28. Gupta, V.; Miura, N. Polyaniline/single-wall carbon nanotube (pani/swcnt) composites for high performance supercapacitors. *Electrochim. Acta* **2006**, *52*, 1721–1726. [CrossRef]
29. Dong, B.; He, B.-L.; Xu, C.-L.; Li, H.-L. Preparation and electrochemical characterization of polyaniline/multi-walled carbon nanotubes composites for supercapacitor. *Mater. Sci. Eng. B* **2007**, *143*, 7–13. [CrossRef]
30. Kumar, M.P.; Kesavan, T.; Kalita, G.; Ragupathy, P.; Narayanan, T.N.; Pattanayak, D.K. On the large capacitance of nitrogen doped graphene derived by a facile route. *RSC Adv.* **2014**, *4*, 38689–38697. [CrossRef]
31. Sohn, J.S.; Patil, U.M.; Kang, S.; Kang, S.; Jun, S.C. Impact of different nanostructures of a pedot decorated 3D multilayered graphene foam by chemical methods on supercapacitive performance. *RSC Adv.* **2015**, *5*, 107864–107871. [CrossRef]

Article

A Hollow-Structured Manganese Oxide Cathode for Stable Zn-MnO_2 Batteries

Xiaotong Guo [1], Jianming Li [2,*], Xu Jin [2], Yehu Han [1], Yue Lin [3], Zhanwu Lei [4], Shiyang Wang [4], Lianjie Qin [1,*], Shuhong Jiao [4] and Ruiguo Cao [4,*]

1 School of Environmental and Materials Engineering, Yantai University, Yantai 264005, China; aguoxiaotong@hotmail.com (X.G.); ahanyehu@hotmail.com (Y.H.)
2 Research Institute of Petroleum Exploration and Development (RIPED), PetroChina, Beijing 100083, China; jinxu@petrochina.com.cn
3 Hefei National Laboratory for physical Science at the Microscale, University of Science and Technology of China, Hefei 230026, China; linyue@ustc.edu.cn
4 Key Laboratory of Materials for Energy Conversion Chinese Academy of Sciences (CAS), Department of Materials Science and Engineering, University of Science and Technology of China, Hefei 230026, China; zwlei@mail.ustc.edu.cn (Z.L.); wangshiy@ustc.edu.cn (S.W.); jiaosh@pku.edu.cn (S.J.)
* Correspondence: lijm02@petrochina.com.cn (J.L.); ljqin@ytu.edu.cn (L.Q.); rgcao@ustc.edu.cn (R.C.); Tel.: +86-10-8359-2647 (J.L.); +86-535-690-2239 (L.Q.); +86-551-6360-1807 (R.C.)

Received: 3 April 2018; Accepted: 27 April 2018; Published: 5 May 2018

Abstract: Aqueous rechargeable zinc-manganese dioxide (Zn-MnO_2) batteries are considered as one of the most promising energy storage devices for large scale-energy storage systems due to their low cost, high safety, and environmental friendliness. However, only a few cathode materials have been demonstrated to achieve stable cycling for aqueous rechargeable Zn-MnO_2 batteries. Here, we report a new material consisting of hollow MnO_2 nanospheres, which can be used for aqueous Zn-MnO_2 batteries. The hollow MnO_2 nanospheres can achieve high specific capacity up to ~405 mAh g^{-1} at 0.5 C. More importantly, the hollow structure of birnessite-type MnO_2 enables long-term cycling stability for the aqueous Zn-MnO_2 batteries. The excellent performance of the hollow MnO_2 nanospheres should be due to their unique structural properties that enable the easy intercalation of zinc ions.

Keywords: manganese oxide; hollow structure; multivalent intercalation; zinc ion batteries

---

1. Introduction

Lithium-ion batteries (LIBs) have predominantly held a significant share of the energy storage market for portable electronics and electric vehicles since the 1990s, due to their high energy/power density and long cycling life. However, with the rapid development of renewable energy plants, there is an extensive and urgent demand for energy storage technologies for large-scale smart grid applications, which require rechargeable battery systems with good cycling performance, low cost, high safety, and environmental friendliness. In searching for new chemistry beyond lithium-ion batteries, multivalent secondary batteries (Mg, Ca, Zn, and Al) have attracted tremendous research efforts, which could, in principle, deliver a higher energy density based on their multi-electron reaction mechanisms [1,2]. Among the multivalent batteries based on intercalation chemistries, aqueous rechargeable zinc ion batteries are considered as a promising candidate for large-scale energy storage applications because of their low cost and the large abundance of Zn [3]. In addition, the aqueous electrolytes in zinc ion batteries provide better safety compared to other battery systems with flammable organic electrolytes. However, the development of aqueous zinc ion batteries is significantly hindered by the limited choice of positive electrode materials, which usually suffer from low specific capacity and poor cycling stability [4]. Many failure mechanisms are associated with phase transformations and the formation of irreversible products [5,6]. Only a few positive electrodes coupled with suitable electrolytes have been demonstrated to be able to achieve stable long-term cycling for aqueous zinc ion batteries [7–12].

Despite their low cost and high abundance, manganese oxides have a variety of advantages including tunable crystal structure and a scalable manufacturing process, which have been widely used for many energy storage applications including lithium-ion batteries, supercapacitors, and zinc-air batteries [13–15]. Manganese oxides possess a variety of polymorphs, including α-, β-, γ-, δ-, λ-, and ε-types, which form different structures such as tunnel, layered, and spinel structures, and can be used as positive electrode materials for aqueous zinc manganese dioxide (Zn-MnO_2) batteries [16–19]. Birnessite-type manganese dioxide (δ-MnO_2) is featured with a layered structure, which is considered as a favorable host for the intercalation of various cations [20,21]. Considerable efforts have been made to verify this layered structure materials for reversible zinc ion intercalation [22]. It was observed that the birnessite-type manganese dioxide is not stable as a positive electrode material under the long-term cycling of a secondary Zn-MnO_2 battery [23]. In order to deliver a two-electron capacity for a long cycling life, the structure of δ-MnO_2 needs to be maintained by structure-stabilizing agents. For example, it was reported that the birnessite-type MnO_2 could achieve a full two-electron capacity for over 6000 cycles when mixed with bismuth oxide (Bi_2O_3), called Bi-birnessite (Bi-δ-MnO_2), intercalated with Cu^{2+} ions [24]. Also, we note that hollow nanostructures offer promising potentials for energy storage applications because of their favorable properties in terms of hierarchical structure complexity and fast ion transport pathway [25,26].

Herein, without stabilizing agents, we tackle the stability issue of δ-MnO_2 in aqueous Zn-MnO_2 batteries by tuning the nanostructure of this materials. A hollow spherical structure of δ-MnO_2 is developed to enable a robust architecture and a high specific capacity of the positive electrode for an aqueous Zn-MnO_2 battery. The hollow manganese oxide cathode exhibits high capacity and stable cycling performance with an aqueous electrolyte.

2. Materials and Methods

2.1. Synthesis of Hollow Spherical MnO_2 Particles

SiO_2 spherical particles were prepared by a sol-gel method and used as a template. In a typical synthesis procedure, 4.0 mL of tetrapropyl orthosilicate was added into the mixture of ethanol (50.0 mL), water (10.0 mL), and ammonia (1.0 mL, 25–28%) at room temperature under stirring. After 14 h, the obtained SiO_2 suspension was centrifuged, rinsed with distilled water, and re-dispersed in 30 mL H_2O to form a SiO_2 white suspension.

Then, 0.98 g of $KMnO_4$ was added to the SiO_2 suspension and followed by ultrasonic treatment for 30 min. The suspension was then transferred to a Teflon-lined autoclave and heated at 150 °C for 48 h. The brown product with a silica/manganese oxide core-shell structure (SiO_2@MnO_2) was obtained and then etched in the 2.00 M of $NaCO_3$ solution at 60 °C for 24 h.

After the removal of the SiO_2 core, the final products of the hollow spherical MnO_2 particles were collected by centrifugation, washed with deionized water, and freeze-dried.

2.2. Cell Assembly and Test

To prepare the cathode electrode, the slurry was prepared with 70 wt % MnO_2, 20 wt % KB (Ketjenblack), and 10 wt % PVDF (Polyvinylidene Fluoride) binder and casted onto a Ti foil current collector. The electrode was dried at 60 °C in a vacuum oven for 24 h. The loading of MnO_2 on the electrodes was around 0.5 mg/cm^2. The CR2032 coin cells were assembled with zinc metal as anodes and MnO_2 as cathodes. The electrolyte was 1.0 M $ZnSO_4$ with 0.2 M $MnSO_4$ as an additive and glass fiber was used as the separator. Galvanostatic measurements were carried out between 1.0 and 1.8 V on a Land CT2001A system (LANHE, Wuhan, China). The cyclic voltammetry (CV) experiments were performed with a CHI600E electrochemical workstation (CH, Shanghai, China) at a scanning rate of

0.1 mV s^{-1} between 0.8 and 1.9 V. The electrochemical impedances spectroscopy (EIS) of the active material was recorded on an electrochemical workstation (Solartron) using the frequency response analysis with a range from 100 kHz to 0.01 Hz.

2.3. Materials Characterization

The dimensions and morphologies were examined using scanning electron microscopy (SEM, JSM-2100F, JEOL, Tokyo, Japan. The crystallographic structures were investigated by powder XRD (X-ray diffraction) measurements on a Rigaku D/max-TTR III diffractometer with Cu Kα radiation (Rigaku Corporation, Shibuya-ku, Japan), 40 kV, 200 mA. The nanostructures of hollow spherical samples were characterized by high-resolution transmission electron microscopy (HRTEM, JEOL, Tokyo, Japan, 2010).

3. Results

The hollow MnO_2 nanospheres were synthesized using a template approach. The synthesis process of hollow MnO_2 nanospheres is illustrated schematically in Figure 1. First, the SiO_2 nanospheres were prepared through a sol-gel method. To form the core-shell structure of SiO_2@MnO_2, the as-synthesized SiO_2 nanospheres were used as templates for a hydrothermal process with a $KMnO_4$ solution. After being etched in an aqueous Na_2CO_3 solution, the SiO_2 core was removed and the hollow MnO_2 nanospheres were obtained for characterization and electrochemical tests.

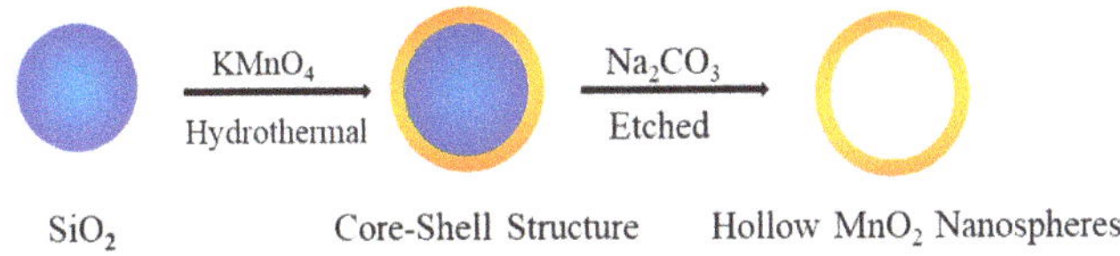

Figure 1. Schematic illustration of the synthetic process of hollow MnO_2 nanospheres.

As shown in Figure 2a, the prepared monodisperse SiO_2 nanospheres show a uniform sphere morphology with a size ranging from 200 to 250 nm. After the reaction with aqueous $KMnO_4$ solution by a hydrothermal process at 150 °C for 48 h, the core-shell structure of SiO_2@MnO_2 was formed (Figure 2b). It was clearly shown that the SiO_2 nanospheres were fully covered with MnO_2 and no aggregation was observed. The uniform coating on SiO_2 nanoparticles was due to the surface-induced nucleation and growth of manganese oxide species. To remove the SiO_2 core materials, the core-shell SiO_2@MnO_2 particles were etched in an aqueous 2 M Na_2CO_3 solution for 24 h. After the etching process, very little silica is remained based on EDX (Energy Dispersive X-Ray Spectroscopy) and XPS (X-ray Photoelectron Spectroscopy) measurements. Figure 2c shows the typical morphology of hollow spherical MnO_2 particles after the etching treatment. It is clearly seen that the spherical morphology is completely maintained and almost no damage was observed on the shell structure of MnO_2. Powder X-ray diffraction (XRD) measurement was used to examine the crystallographic structure phase in the as-synthesized hollow MnO_2 spheres. Figure 2d shows the XRD pattern of the as-synthesized hollow MnO_2 nanospheres, which shows peaks at 2θ around 12.4°, 24.8°, 36.8°, and 65.8°. These peaks can be indexed to birnessite-type MnO_2. The peaks lack the long-range order of layers and a tail toward higher angle two-theta, demonstrating common features of the birnessite structure [27].

In order to further investigate the structure of the as-synthesized hollow MnO_2 nanospheres, we carried out high-resolution TEM analysis. Figure 3a clearly shows the hollow structure of MnO_2 nanospheres without aggregation observed. The MnO_2 shell is around 15 nm thick and its diameter is around 200 nm. Almost no damage was observed under TEM analysis, indicating that the shell structure is robust enough to tolerate the harsh etching process. Detailed analysis shows that the shell structure consists of very thin nanosheets of MnO_2, which form interconnected wrinkle structures

(Figure 3b). The wrinkled structure was confirmed by HAADF-STEM (High-Angle Annular Dark Field Scanning Transmission Electron Microscopy) image (Figure 3c). Moreover, elemental compositions of the hollow MnO_2 structure were mapped by electron energy loss spectroscopy (EELS), confirming the uniform dispersion of elemental Mn and O (Figure 3d,e). A N_2 adsorption/desorption analysis of hollow MnO_2 nanospheres was conducted to analyze the surface area of the wrinkled hollow structure. The BET (Brunauer-Emmett-Teller) surface area of as-synthesized hollow MnO_2 nanosphere was ~200 m^2/g with a pore size distribution at ~1.6 nm (Figure 4), indicating that the hollow MnO_2 nanosphere also featured a microporous structure.

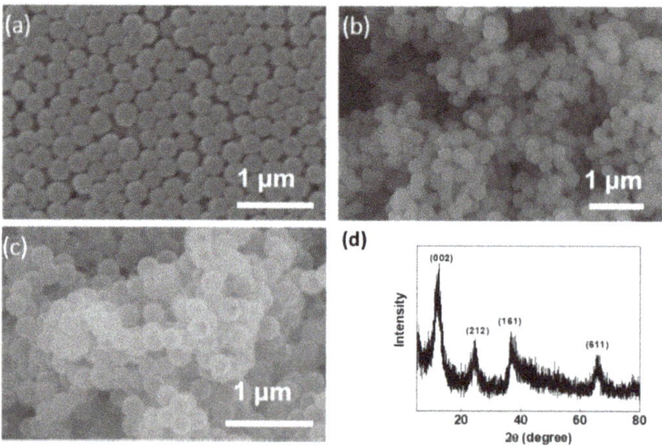

Figure 2. SEM images of SiO_2 nanospheres (**a**); SiO_2@MnO_2 core-shell structure (**b**); and hollow MnO_2 nanospheres (**c**); (**d**) XRD patterns of the hollow MnO_2 nanospheres.

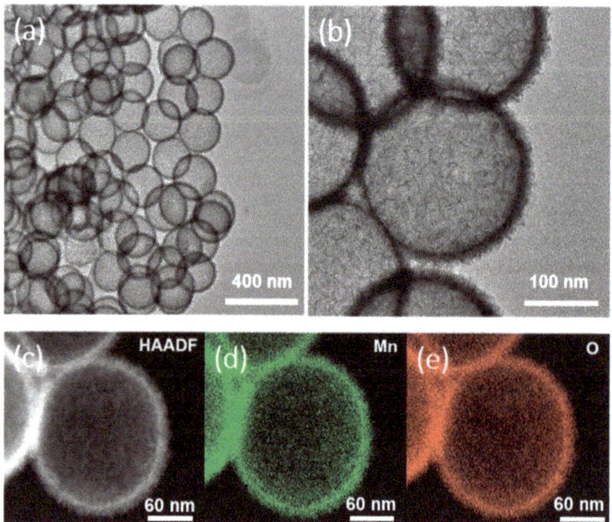

Figure 3. High- (**a**) and low-magnification (**b**) HRTEM images of hollow MnO_2 nanospheres; (**c**) HAADF-STEM image of hollow MnO_2 nanospheres. Elemental mapping of hollow MnO_2 nanospheres: (**d**) Mn and (**e**) O.

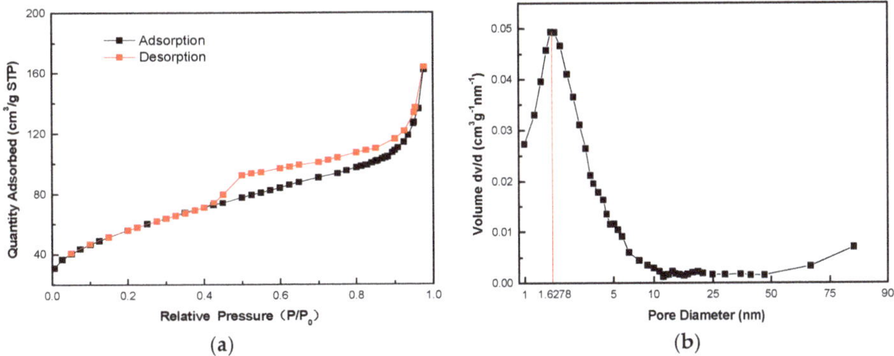

Figure 4. BET measurement of hollow MnO_2 nanospheres. (**a**) Nitrogen adsorption/desorption isotherms of as-synthesized hollow MnO_2 nanospheres; (**b**) the pore size distribution of hollow MnO_2 nanospheres, as calculated using a BJH (Barrett-Joyner-Halenda) method.

The electrochemical performance of hollow MnO_2 nanospheres was evaluated in aqueous Zn-MnO_2 batteries. The Zn-MnO_2 cell was assembled with zinc foil as an anode and 1.0 M $Zn(SO_4)_2$ aqueous solution with 0.2 M $MnSO_4$ as an electrolyte. Figure 5a shows the cyclic voltammetry scan results of the Zn-MnO_2 cell with hollow MnO_2 nanospheres as cathode materials. The sweep range was between 1.9 V and 0.8 V vs. Zn/Zn^{2+}, and the sweep rate was 0.1 mV/s. During the first cycle, a low cathodic peak at around 1.36 V and a sharp cathodic peak at around 1.22 V were observed, while only one anodic peak at around 1.58 V was observed when sweeping back. In the following scan cycles, the cathodic peak at 1.36 V increased gradually, indicating an activation process of hollow MnO_2 nanospheres during discharge. Figure 5b shows the typical galvanostatic discharge/charge profiles of the Zn-MnO_2 cell at a 1 C rate. The discharge curve in first cycle exhibited a flat plateau at around 1.26 V, which is consist with the CV results. Two plateaus, ~1.38 V and ~1.26 V, were observed during the second discharge process, which are related to two distinct cathodic peaks in the second sweep of CV curves, indicating a two-step intercalation process of zinc ions into the birnessite structure. Upon charge process, two plateaus at ~1.50 V and ~1.58 V were observed. Previously, the two-step intercalation process was also observed in other Zn-MnO_2 batteries based on birnessite-type materials [23,24]. The discharge capacity of hollow MnO_2 nanospheres reached up to ~270 mAh g^{-1} at a 1 C rate.

Figure 5c shows the typical charge/discharge profiles of Zn-MnO_2 batteries at different current densities. At rates of 0.5, 1, 2, 5, and 10 C, specific discharge capacities of ~405, ~265, ~166, ~85, and ~40 mAh g^{-1} were obtained, respectively, indicating a good rate performance of the hollow MnO_2 nanospheres. The long-term cycling performance of the Zn-MnO_2 batteries in terms of discharge capacity and coulombic efficiency was also investigated at 1 C. As shown in Figure 5d, we compared the cycling performance of nanosheets, nanorods, and hollow spherical structure of MnO_2 in aqueous Zn-MnO_2 batteries. The morphologies of MnO_2 nanosheets and nanorods are shown in Figure 6. The initial discharge capacity for hollow MnO_2 nanospheres was ~168 mAh g^{-1}. After the activation process, the discharge capacity of the second cycle was reached at ~270 mAh g^{-1}. Notably, after 100 cycles, the discharge capacity was stabilized at ~305 mAh g^{-1} with a coulombic efficiency over 97%. However, the MnO_2 nanorods showed a quickly fading capacity. The MnO_2 nanosheets performed a low discharge capacity and poor cycling performance. The excellent rate capability and cycling stability of the Zn-MnO_2 cell should be due to the hollow structure of the birnessite-type MnO_2 cathode materials.

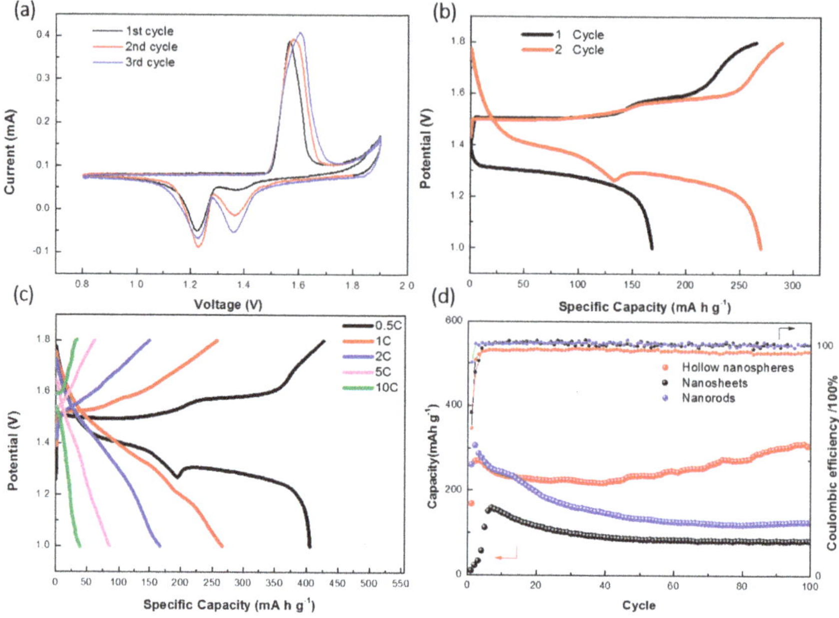

Figure 5. Electrochemical performance of Zn-MnO_2 batteries: (**a**) CV profiles; (**b**) typical charge–discharge curves; (**c**) rate performance; and (**d**) long-term cycling stability of hollow MnO_2 nanospheres, MnO_2 nanosheets, and MnO_2 nanorods at 1 C with an electrolyte of 1.0 M $Zn(SO_4)_2$ and 0.2 M $MnSO_4$.

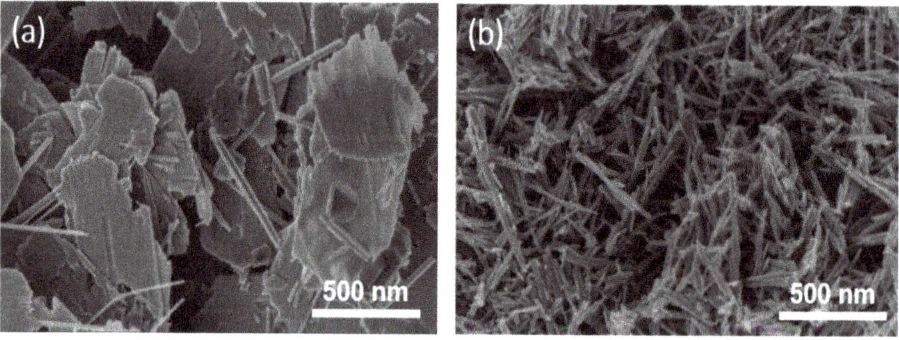

Figure 6. SEM images of (**a**) MnO_2 nanosheets and (**b**) MnO_2 nanorods.

EIS measurements were performed to evaluate the impedance difference between before cycle and after the first discharge/charge cycle. As depicted in Figure 7a, the charge transfer impedance decreased after the first cycle, which indicated that the intercalation of Zn^{2+} ions into the MnO_2 structure became easier after the structure transformation. An ex situ XRD analysis was conducted for the cathode after the first cycle. As shown in Figure 7b, the representative birnessite structure peaks, (002) and (212), significantly decreased in intensity, especially compared to the mixed indices (161) peak. This selective loss suggests a loss of long-range order in the direction of the layers, perhaps due to a structural transformation to another polymorph with similar building blocks but not layered.

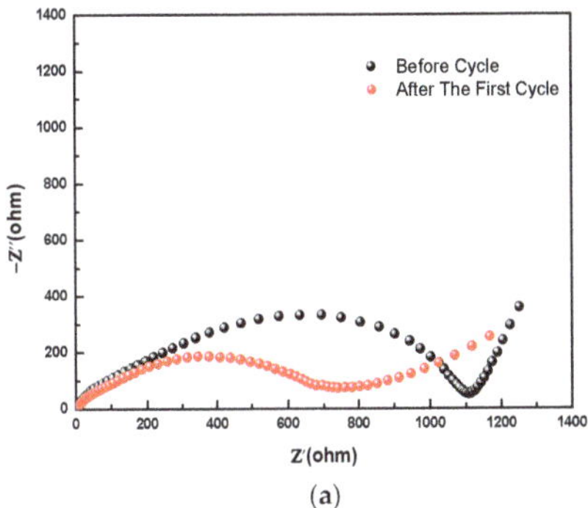

(a)

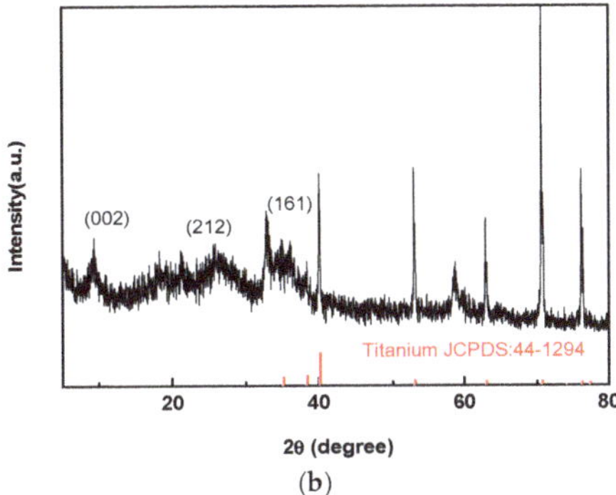

(b)

Figure 7. (**a**) Electrochemical impedance spectra of the Zn/MnO_2 cells before any cycles and after the first cycle; (**b**) XRD patterns of the cathode after the first cycle.

4. Conclusions

In summary, hollow MnO_2 nanospheres were synthesized through a facile hydrothermal approach and used as cathode materials in aqueous Zn-MnO_2 batteries. The hollow birnessite-type MnO_2 cathode achieved a relatively high discharge capacity and stable cycling performance with an aqueous electrolyte in a Zn-MnO_2 battery. The excellent electrochemical performance was ascribed to the unique hollow structure, which favors the intercalation process of zinc ions and enables a stable cycling of the Zn-MnO_2 battery.

Author Contributions: R.C. conceived and designed the experiments; X.G., Y.H., Z.L., and S.W. performed most of the experiments; L.Q. and S.J. analyzed the data; J.L., X.J., and Y.L. contributed reagents/materials/analysis tools; X.G. and R.C. wrote the paper.

Funding: This work was funded by the National Key Research and Development Program of China (Nos. 2017YFA0402800, 2017YFA0206700), the National Natural Science Foundation of China (No. 21776265), and the R&D Department of PetroChina.

Conflicts of Interest: The authors declare no conflict of interest.

References

1. Cheng, Y.W.; Shao, Y.Y.; Raju, V.; Ji, X.L.; Mehdi, B.L.; Han, K.S.; Engelhard, M.H.; Li, G.S.; Browning, N.D.; Mueller, K.T.; et al. Molecular storage of Mg ions with vanadium oxide nanoclusters. *Adv. Funct. Mater.* **2016**, *26*, 3446–3453. [CrossRef]
2. Wang, Y.; Chen, R.; Chen, T.; Lv, H.; Zhu, G.; Ma, L.; Wang, C.; Jin, Z.; Liu, J. Emerging non-lithium ion batteries. *Energy Storage Mater.* **2016**, *4*, 103–129. [CrossRef]
3. Wang, M.; Zhang, F.; Lee, C.S.; Tang, Y.B. Low-cost metallic anode materials for high performance rechargeable batteries. *Adv. Energy Mater.* **2017**, *7*. [CrossRef]
4. Alfaruqi, M.H.; Islam, S.; Mathew, V.; Song, J.; Kim, S.; Tung, D.P.; Jo, J.; Kim, S.; Baboo, J.P.; Xiu, Z.; et al. Ambient redox synthesis of vanadium-doped manganese dioxide nanoparticles and their enhanced zinc storage properties. *Appl. Surf. Sci.* **2017**, *404*, 435–442. [CrossRef]
5. Hertzberg, B.J.; Huang, A.; Hsieh, A.; Chamoun, M.; Davies, G.; Seo, J.K.; Zhong, Z.; Croft, M.; Erdonmez, C.; Meng, Y.S.; et al. Effect of multiple cation electrolyte mixtures on rechargeable Zn-MnO_2 alkaline battery. *Chem. Mater.* **2016**, *28*, 4536–4545. [CrossRef]
6. Alfaruqi, M.H.; Mathew, V.; Gim, J.; Kim, S.; Song, J.; Baboo, J.P.; Choi, S.H.; Kim, J. Electrochemically induced structural transformation in a γ-MnO_2 cathode of a high capacity zinc-ion battery system. *Chem. Mater.* **2015**, *27*, 3609–3620. [CrossRef]
7. Kundu, D.; Adams, B.D.; Ort, V.D.; Vajargah, S.H.; Nazar, L.F. A high-capacity and long-life aqueous rechargeable zinc battery using a metal oxide intercalation cathode. *Nat. Energy* **2016**, *1*. [CrossRef]

8. Zeng, Y.X.; Zhang, X.Y.; Meng, Y.; Yu, M.H.; Yi, J.N.; Wu, Y.Q.; Lu, X.H.; Tong, Y.X. Achieving ultrahigh energy density and long durability in a flexible rechargeable quasi-solid-state Zn-MnO_2 battery. *Adv. Mater.* **2017**, *29*. [CrossRef] [PubMed]
9. Sambandam, B.; Soundharrajan, V.; Kim, S.; Alfaruqi, M.H.; Jo, J.; Kim, S.; Mathew, V.; Sun, Y.K.; Kim, J. Aqueous rechargeable Zn-ion batteries: An imperishable and high-energy $Zn_2V_2O_7$ nanowire cathode through intercalation regulation. *J. Mater. Chem. A* **2018**, *6*, 3850–3856. [CrossRef]
10. Zhang, N.; Cheng, F.Y.; Liu, Y.C.; Zhao, Q.; Lei, K.X.; Chen, C.C.; Liu, X.S.; Chen, J. Cation-deficient spinel $ZnMn_2O_4$ cathode in $Zn(CF_3SO_3)_2$ electrolyte for rechargeable aqueous Zn-Ion battery. *J. Am. Chem. Soc.* **2016**, *138*, 12894–12901. [CrossRef] [PubMed]
11. Cheng, Y.W.; Luo, L.L.; Zhong, L.; Chen, J.Z.; Li, B.; Wang, W.; Mao, S.X.; Wang, C.M.; Sprenkle, V.L.; Li, G.S.; et al. Highly reversible zinc-ion intercalation into chevrel phase MO_6S_8 nanocubes and applications for advanced zinc-ion batteries. *ACS Appl. Mater. Interfaces* **2016**, *8*, 13673–13677. [CrossRef] [PubMed]
12. Hu, P.; Zhu, T.; Wang, X.; Wei, X.; Yan, M.; Li, J.; Luo, W.; Yang, W.; Zhang, W.; Zhou, L.; et al. Highly durable $Na_2V_6O_{16}\cdot 1.63H_2O$ nanowire cathode for aqueous zinc-ion battery. *Nano Lett.* **2018**, *18*, 1758–1763. [CrossRef] [PubMed]
13. Tang, Y.; Zheng, S.; Xu, Y.; Xiao, X.; Xue, H.; Pang, H. Advanced batteries based on manganese dioxide and its composites. *Energy Storage Mater.* **2018**, *12*, 284–309. [CrossRef]
14. Hou, Y.; Cheng, Y.; Hobson, T.; Liu, J. Design and synthesis of hierarchical MnO_2 nanospheres/carbon nanotubes/conducting polymer ternary composite for high performance electrochemical electrodes. *Nano Lett.* **2010**, *10*, 2727–2733. [CrossRef] [PubMed]
15. Zhang, K.; Han, X.P.; Hu, Z.; Zhang, X.L.; Tao, Z.L.; Chen, J. Nanostructured Mn-based oxides for electrochemical energy storage and conversion. *Chem. Soc. Rev.* **2015**, *44*, 699–728. [CrossRef] [PubMed]
16. Islam, S.; Alfaruqi, M.H.; Song, J.; Kim, S.; Pham, D.T.; Jo, J.; Kim, S.; Mathew, V.; Baboo, J.P.; Xiu, Z.; et al. Carbon-coated manganese dioxide nanoparticles and their enhanced electrochemical properties for zinc-ion battery applications. *J. Energy Chem.* **2017**, *26*, 815–819. [CrossRef]
17. Ko, J.S.; Sassin, M.B.; Parker, J.F.; Rolison, D.R.; Long, J.W. Combining battery-like and pseudocapacitive charge storage in 3D MnO_x@Carbon electrode architectures for zinc-ion cells. *Sustain. Energy Fuels* **2018**, *2*, 626–636. [CrossRef]
18. Xu, C.J.; Li, B.H.; Du, H.D.; Kang, F.Y. Energetic Zinc ion chemistry: The rechargeable Zinc ion battery. *Angew. Chem.-Int. Ed.* **2012**, *51*, 933–935. [CrossRef] [PubMed]
19. Zhang, N.; Cheng, F.Y.; Liu, J.X.; Wang, L.B.; Long, X.H.; Liu, X.S.; Li, F.J.; Chen, J. Rechargeable aqueous Zinc-Manganese dioxide batteries with high energy and power densities. *Nat. Commun.* **2017**, *8*. [CrossRef] [PubMed]
20. Sun, X.Q.; Duffort, V.; Mehdi, B.L.; Browning, N.D.; Nazar, L.F. Investigation of the mechanism of mg insertion in birnessite in nonaqueous and aqueous rechargeable mg-ion batteries. *Chem. Mater.* **2016**, *28*, 534–542. [CrossRef]
21. Nam, K.W.; Kim, S.; Lee, S.; Salama, M.; Shterenberg, I.; Gofer, Y.; Kim, J.S.; Yang, E.; Park, C.S.; Kim, J.S.; et al. The high performance of crystal water containing manganese birnessite cathodes for magnesium batteries. *Nano Lett.* **2015**, *15*, 4071–4079. [CrossRef] [PubMed]
22. Han, S.D.; Kim, S.; Li, D.G.; Petkov, V.; Yoo, H.D.; Phillips, P.J.; Wang, H.; Kim, J.J.; More, K.L.; Key, B.; et al. Mechanism of Zn insertion into nanostructured delta-MnO_2: A nonaqueous rechargeable Zn metal battery. *Chem. Mater.* **2017**, *29*, 4874–4884. [CrossRef]
23. Alfaruqi, M.H.; Gim, J.; Kim, S.; Song, J.; Pham, D.T.; Jo, J.; Xiu, Z.; Mathew, V.; Kim, J. A layered delta-MnO_2 nanoflake cathode with high Zinc-storage capacities for eco-friendly battery applications. *Electrochem. Commun.* **2015**, *60*, 121–125. [CrossRef]
24. Yadav, G.G.; Gallaway, J.W.; Turney, D.E.; Nyce, M.; Huang, J.C.; Wei, X.; Banerjee, S. Regenerable Cu-intercalated MnO_2 layered cathode for highly cyclable energy dense batteries. *Nat. Commun.* **2017**, *8*. [CrossRef] [PubMed]
25. Yu, L.; Hu, H.; Wu, H.B.; Lou, X.W. Complex hollow nanostructures: Synthesis and energy-related applications. *Adv. Mater.* **2017**, *29*. [CrossRef] [PubMed]

26. Zhou, L.; Zhuang, Z.C.; Zhao, H.H.; Lin, M.T.; Zhao, D.Y.; Mai, L.Q. Intricate hollow structures: Controlled synthesis and applications in energy storage and conversion. *Adv. Mater.* **2017**, *29*. [CrossRef] [PubMed]
27. Birkner, N.; Navrotsky, A. Thermodynamics of manganese oxides: Sodium, potassium, and calcium birnessite and cryptomelane. *Proc. Natl. Acad. Sci. USA* **2017**, *114*, E1046–E1053. [CrossRef] [PubMed]

MDPI
St. Alban-Anlage 66
4052 Basel
Switzerland
Tel. +41 61 683 77 34
Fax +41 61 302 89 18
www.mdpi.com

Nanomaterials Editorial Office
E-mail: nanomaterials@mdpi.com
www.mdpi.com/journal/nanomaterials

www.ingramcontent.com/pod-product-compliance
Lightning Source LLC
LaVergne TN
LVHW070626170726
843515LV00004B/187

9783039362783